First Edition 2021
© Paul B. Martin
chiringapress@gmail.com

ISBN 978-1-61012-046-3

> "Dharmaists", "Moses", Siddhartha,
> Paul, Muhammad, Erasmus, (or the
> women around them) stressed:
> Simplicity and Humility; Caring and Sharing;
> PEACE...as TRUTHS.

GAMES WE PLAY

MORE THAN 200,000 YEARS
OF LIVING TRUTHFULLY
AND 5,000 OF SEARCHING
FOR TRUTH

BY AND FOR
ALL BIOTA

An abbreviated book with abundant illustrations all about
applied ecology *for kids from 12 years of age to 120.*

Seguin, Texas

Dedicated to

Mom St. Louise Katherine (Kneuper) Martin

and

Dad Luther Alton Martin

ACKNOWLEDGEMENTS

In the introductions and other sections in this book, I, paul bain martin, have mentioned many who have had a positive influence on me in my seventy plus years. Nevertheless, I would also like to express particular appreciation to the following special individuals for opening up exciting and gratifying ecological pathways taken in my life:

Mom Louise and Dad Alton Martin for a foundation for living *sabiamente*, simply, smally, slowly, steadfastly, sharingly, and sustainably and for living in solidarity with the poor, powerless, and disenfranchised.

Uncle L.C. "Peggy" Martin, Texas agriculture entrepreneur, for introducing me to agricultural entomology as an exciting career.

Dr. Joseph Schaffner, Texas A&M University, for giving me real purpose and stimulating me toward transitioning to biology as a life pursuit and sustainable livelihood.

Dr. Richard L. Ridgway, ARS, USDA, for immersing me in biocontrol, population dynamics, and applied ecology.

Elizabeth Florence (Hoffmann) Martin of Rio Medina for life-long support as my wife, wonderful friend and colleague, and true love.

Dr. Peter D. Lingren, ARS, USDA, for his undying support and his insistence that I persevere in pursuing my dreams despite a lack of certain talents.

Dr. David Pimentel, Cornell University, for an essential ecological foundation in energetics.

Dr. Miguel A. Altieri, University of California, Berkeley, for prodding me to do what I know is right.

Dr. Lanier E. Byrd, St. Philip's College, for providing me and my family a sustainable livelihood in community at a wonderful Historically Black, Hispanic-Serving Institution for twenty-plus years.

Humankind ... and Truth

Humankind.
Kinds of humans.
Kind humans?

Whether relatively wealthy or very poor,
Fossil-fuel greased or with calloused hands,
... Most, deservedly or greedily, seek MORE!

And even with promises made to tithe and do with less
We all generally continue to clutch power possessed
... Within our abominable mess.

Food, fiber, and shelter from, or in, far-away places.
Expansive homes; extravagant, landscaped spaces.
Land with too many cleared, paved, asphalted, concreted lanes.
Eaarth[1] with far too many automobiles and planes,
And PCs and androids,
And loud electronic and mechanical noise.
....
Elegant poems, real poetry,
Non-algorithms of ancient and contemporary times,
Can be—perhaps should be—
Succinct and short?
Let me try ... toward truth—

Ali's "Me, We."
Keats' "Beauty is truth, truth beauty.
That is all ye know on earth,
And all ye need to know."
But perhaps with lots of Thoreau?
But what—then—is me, and who are we?

---

1. Term used by ecological activist Bill McKibben for Earth of the Anthropocene.

And is there actually more to beauty? And more to truth?
The anti-algorithms,
Especially if elegant and real—
Succinct, short—
Prod us to ponder.

Overshoot? Rape and exploitation? Disparity?
Are not these, truthfully, challenges to real beauty?
For me? For we?

Life that is prudent,
Simple, humble, and sharing.
Solidarity!
Positively Ethical
 ... Applied Community Ecology.
PEACE.
That should be me.
That IS "We!"
It is beauty.
It is truth.

Introduction

In the following pages of this book you can browse through and hopefully be compelled to study, a glossary, some illustrations and outlines, and brief stories, poems, and essays. Each entry was designed to stand alone but also to connect toward an understanding of what the undertaking of positively ethical applied community ecology, or **PEACE**, means. We wish to hook the reader into this PEACE journey of seeking more knowledge, becoming wiser, and entering into an ethical and moral livelihood and life of profound and holistic prudence.

Herein we are trying to:

- make a case for **Why?** we should live sustainably, i.e., through a hard-fought process of social justice, humaneness, and ecological sanity,

- point out **What?** we need to do in order to realize resilient, sustainable ecological community, and

- suggest **How?** we can realize the process and the Whats.

We hope this book is helpful toward realizing the following:

- the 7 S's, or living *sabiamente*/**wisely, simply, smally, slowly, steadfastly, sharingly, sustainably,**
- positively ethical applied community ecology,
- regeneration and conservation of resilient, sustainable community,
- transformation of Eaarth back toward being more of an Earth, and
- a healthier world for all for as long as possible.

(The introduction, including the glossary with definitions for new and old words and phrases, text, and appendix has been developed by paul bain martin. Illustrations are primarily by Laura Salazar, with help from daughter Joaquina Guevara, and Elizabeth Florence Martin.)

Limits. I do wish to explain here from the get-go how we determined the somewhat arbitrary proposed upper boundary lines (goals) for energy transformation (70,000 kilocalories per capita daily), ecological footprints (5 acres per capita), and earnings per capita yearly in 2020 U.S. dollars ($50,000), i.e., power over resources.

1. There are limits to the natural resource base and to what the evolved living systems of homeostasis can take before totally unraveling. Too much energy transformation and corresponding excessive ecological footprints will try those limits. In keeping with the Precautionary Principle, we need to educate a human population of almost 8 billion (heading to 10 billion) toward regulating consumption and imprudent energy transformation and ecological footprints. Moreover, morally and ethically we need to have goals of equity and equality.

2. Some of the reviewers of this book have stated that we should go easier on overtly "beating up" on us Haves and put much more emphasis on bright hope. However, when we take a hard look at the data worldwide, we immediately get despondent and depressed and anxious for immediate change. There are so very many who are making do with so little because a few (percentwise) are not sharing and are squandering and destroying so much. (There are many sources for these data. In terms of energetics, the hard-working scientist and prolific author, Vaclav Smil is one of the best sources. And the World Wildlife Fund has done an amazing job with its regular Living Planet Reports.)

3. Our suggested limits to energy transformation, ecological footprints, and yearly earning are actually too high!

In this current time the average human is transforming about 39,000 kilocalories per capita daily. By 2050, even if that average for energy transformation remains stable, the total transformation of energy will increase by about 1.7 times because of population increase from 7-8 billion to ca. 10 billion. Nevertheless, in our rough calculations for proposed boundaries/goals, we did almost double the current level for a goal of ca. 70,000[1] kilocalories per capita daily as a limit. This would mean that the we Haves of the U.S. would need to cut their consumption from an average of ca. 200,000 by about 2/3rds. (We have proposed ecological footprints of 5 global acres per capita which is "slightly" over the Eaarth's biocapacity and considerably less that the U.S. footprint of about 17 global acres.)

For power over resources, i.e., per capita income per year, we selected the recent annual income for the average U.S. citizen. This is definitely too high. We do wish to underline the point that humans in other parts of the world deserve more power and equity and that the U.S. should not have a monopoly on that power.

(Smil, Vaclav. 2000. Energy in the twentieth century: Resources, conversions, costs, uses, and consequences. Annu. Rev. Energy Environment. 25: 21-51

2019 Revision of World Prospects. Dep. Of Economic and Social Affairs. Population Dynamics. United Nations

Living Planet Report, 2016, World Wildlife Fund

State of the States. Global Footprint Network and Earth Economics. July 2015.)

Global Climate Change. Even though we did not mention global climate change much herein, we do believe it is a serious threat to dynamic homeostatic symbioses/"nature" and quality life for all for as long as possible. Moreover, if we would take actions toward living the 7 S's, or *sabiamente*, simply, smally, slowly, steadfastly, sharingly, and sustainably this view will address climate change as well as other ecological challenges.

Socialism vs. Capitalism. We do believe it is important to discuss the merits and downsides of various socio-political/economic systems in polemic as well as in calm diplomatic processes and that these actions should involve critical thinking and decision-making. Suffice it to say at this moment that human organizations for governing are called governments, and these entities are especially important for a civil society and world. We need good government to guide the "invisible hand of

---

1. From my first napkin calculations I proposed 60,000 kilocalories per capita daily as a boundary/goal and have this in many of my writings. Moreover, at times I have proposed 7 global hectares as a boundary/goal for an average ecological footprint. … Again, both of these are too high given the Earth's biocapacity and in considering dynamic homeostatic symbioses, but they are considerably lower than the numbers for the average citizen of the U.S.

capitalism" toward equity, fairness, and justice for all and to discourage de facto slavery and usury, and eliminate genocide and exploitation of Land and the powerless, including other species. And yes, we can use Democratic Socialism as a term for a collective governmental approach to management of neoliberal or laissez-faire capitalism. Democratic Socialism has the potential to be exceptionally good despite its smearing by certain sectors of the populace.

COVID-19. Finally, we are a social animal which works naturally in community toward dynamic homeostatic symbioses. The COVID-19 pandemic and various local responses globally have demonstrated that it is possible for government to help facilitate maintenance of this process of homeostatic symbiosis. It is difficult, but if we put our collective minds to it, we can live *sabiamente*/wisely, simply, smally, slowly, steadfastly, sharingly, SUSTAINABLY.

Table of Contents

> Why don't you humans practice what you preach?
> ... the Golden Rule ... PEACE ... Love ... Humility.

What Drove Us To Develop This Book!

The sky is falling for some living things at various points in time in the universes and cosmos. This can be observed through the lens of the Hubble space telescope, at eye levels here on Earth, or in the use of micro & nano technologies. On the other hand, in all these dimensions, in certain spatial/temporal sectors for some species complexes (ecological communities), life is good! Moreover, this dynamism, incongruity, dilemma, enigma, mystery of humanity, Nature,[1] and Eaarth[1]—which includes war and catastrophes as well as lovers' love and beautiful harmonies--can be studied and to some extent understood via physics, chemistry, geology, paleontology, anthropology, sociology, and psychology and/or a combined application-process of all of these and other disciplines in a manner which profoundly and comprehensively focuses on life and living systems, i.e., biology and ecology.

Despite the enigmatic, dark, and fearful mysteries, and probably because of them, we all love life. We might especially love life, for example, during a day in my place of residence, Clean (or Real) Seguin, Texas, U.S.A., when we have the grandkids over and it is a cool fall day after a recent rain or when the birds are chirping and we're sipping on a *caipirinha* made with quality *cachaça*, partaking of some of cousin Gilda Colley's crunchy, tasty strudel, and we're listening to John Prine singing "You Got Gold" (or even "Some Humans Ain't Human").

We who collaborated in developing this book do love life, and especially children and grandchildren! But beyond offspring, life includes all species, and the authors seek quality of life and healthy life systems for

---

1. Words like this and associated terms which obviously need to be clarified and/or defined for many readers are defined and even discussed in the glossary.

ALL, and for as long as possible. We are, as Edward O. Wilson proposes, WE ARE holistically biophilic.

Furthermore, in order to realize this love and to have healthy, quality lives for all, we recognize the urgent need for comprehensive, in-depth education or ecology across curricula and campuses of all human organizational entities. The appropriate application of this knowledge will inevitably involve a process of reducing growth of human and domestic animal numbers, a reduction of consumption and collective ecological footprints and redistribution of power to powerless humans, and other species. Herein we are using the symbol of: 7 S's / VV->^^, and the phrase, positively ethical applied community ecology/PEACE, for this very appropriate journey.

There were four major collaborators on this book dealing with applied ecology. A significant force, Joaquina Guevara, a Gen-Next and the youngest involved, received her ecological ethos from her generalist and agitator of a mother, Laura Salazar, and from a boarding school and at the liberal arts school of Wesleyan University in the northeastern U.S. Laura, a Gen-X, whose mother was from Guanajuato, Mexico, became a woman of many ecological talents, including the arts, in a hard-scrabble world in the *barrios* of Houston and San Antonio.

A key collaborator on this little book, paul bain martin, is convinced that if everyone in the world lived sharing lives of small ecological footprints, such as that of his parents, Luther Alton Martin & Louise Katherine (Kneuper) Martin, the world would be relatively sustainable and much better in terms of quality life for all. During his formative years paul was raised by Alton and Louise, or rather Louise and Alton, in a family of eight on a five-acre diversified hog farm in a two-bedroom home. After Alton booted him out to sort of fly ecologically on his own, he earned degrees in agricultural entomology from Texas A&M and the University of Florida. He started adding to what he learned about applied ecology from Louise and Alton, Ms. Ruth Allen, Mrs. D. Marshall, and Henry Moss and others in Devine, Texas, and environs, with more academic learning facilitated and taught through direct guidance and/or books of R.L. Ridgway, E.J. Dyksterhuis, P.D. Lingren, D. Pimentel, Archie Carr, H.T. Odum, E. Farber, V. Smil, H. Haberl, and others. Later Miguel Angel Altieri, P. Sechrist, P. Maddox, D. Birkenfeld, Lupe Romero Ramsey, Rosa Lilliam Gomez Diaz, Marvel Maddox, and Tim Barr helped push him into applied ecology efforts which involved advocacy and activism. Others who very significantly inspired paul have been Dr. Joseph Schaffner, E.O. Wilson, D. Suzuki, Wendell Berry, D. Orr, J. Diamond, W. Jackson, D. Worster, V. Prashad, F. Kirschenmannn, C. Miller, and J. Kiel. Recently Hilario Martinez, Susan Kinne, Alphonso Rincon, and Helene and Waylon Gaddie, have been wonderful role models.[2]

We must acknowledge the contributions of Sylvia Manning and Seguin, Texas' Creekside Poets who were the stimuli for many of the poems herein and Gerry Richardson who labored in editing a final manuscript. Tim Barr invested considerable time on the original draft in ca. 2017 and suggested that stories and other narrative be included with what was originally only illustrations. Moreover, Ana María González and Michael Godeck provided wonderful encouragement and guidance to the final publication of this material.

However, it was paul's wife Betsy, Elizabeth Florence Hoffmann Martin, and his children, their spouses, and grandkids who have sacrificed most for paul. They have given him a major impetus in his efforts toward PEACE. Moreover, son John Alton was particularly essential with his constant help with computer soft and hardware and his patience in helping baby his parents through this world of appropriate and inappropriate technology.

paul and Betsy's lives began immediately after World War II in the baby boom of conventional capitalism and massive advertising propaganda, consumerism, and materialism. However, their parents, influenced greatly by the Great Depression and a simple rural ethic, were humble, cautious, and frugal, and demonstrated that life can be full of love, learning, and robust community interactions while being *sabido*, simple, small, slow, steadfast, sharing, sustainable. Moreover, Catholic schoolings, teachings, and preachings infused in paul and Betsy a desired lifestyle of the 7 S's. Furthermore, the civil rights-, anti-war-, and holistic ecological health-movements of the 1960's and 70's expanded a personal desired ethos of empathizing and sharing with other humans and other species, i.e., an Ethic of Reciprocity, or a holistic interpretation and realization of the Golden Rule.

---

2. There were also many others, e.g., Pope John the 23rd, M.L. King, Jr., Cesar Chavez, as well as leaders/teachers/coaches in pbm's hometown Devine-Texas. We do wish that we could acknowledge and list them all.

Even though paul, in particular, has always had trouble articulating eco-literacy and eco-values (as is the case with most folk), he eventually began to passionately believe that rather than seeking individual and tribal power, money, stuff, glamour, arrogant satisfaction ...

1. We need to fervently and massively work toward reducing the individual and collective ecological footprints of the Haves (perhaps 0.5-1 billion humans, ca. 2020) and living the 7 S's. (Reduction should be from about 150,000-300,000 kilocalories used/capita/day to about 70,000, or a reduction of about 2/3rds. In acres the ecological footprint in the U.S.A. should be reduced from about 17 global acres/capita to five (5). ... Also, growth of human and domesticated animal population numbers needs to be reduced.

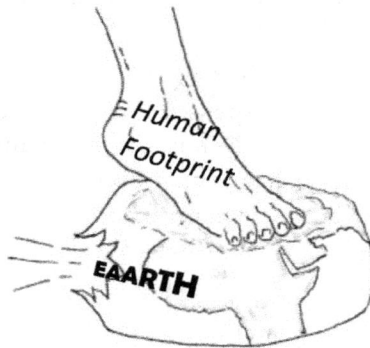

2. The power over the natural resource base (top soil and quality air, quality water, photosynthesizers, biodiversity, free available energy) needs to be shared and shifted toward the 3 billion (ca. 2020) have-nots, and especially to the 0.5 to 1 billion in extreme poverty, and to other species (Ethic of Reciprocity).

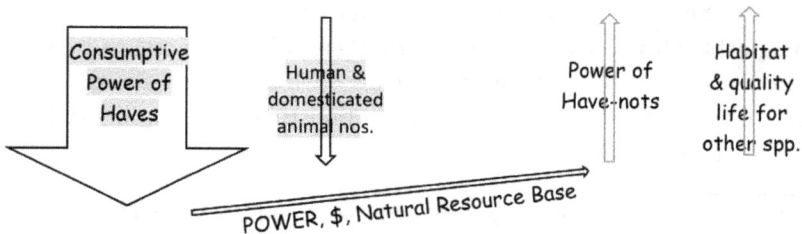

3. Major actions and progress toward a world of the 7 S's would include:

- Realization of positively ethical applied community ecology-PEACE across curricula and campuses of all human organizational entities. (All parents/adults should be tasked with teaching their children/other kids … biology and PEACE.)

- Open borders (but with regulation and caution in realizing this in order to not have net harm to have-not humans and other species).

- Setting aside ½ of Eaarth (with a goal of it becoming Earth again) to Nature *(Proposed by E.O. Wilson et al.; Wilson 2016 Half-Earth: Our Planet's Fight for Life).*

• Realizing appropriate applied agroecology *(As proposed by Dr. M.A. Altieri et al.; Altieri 1989 Agroecology: The Science of Sustainable Agriculture).*

Floresta

• Having a significant (critical) mass of humans from everywhere who would go to war zones, and actively, but peaceably!!/non-violently!, protest actions of war and the possession and use of armaments/weapons.

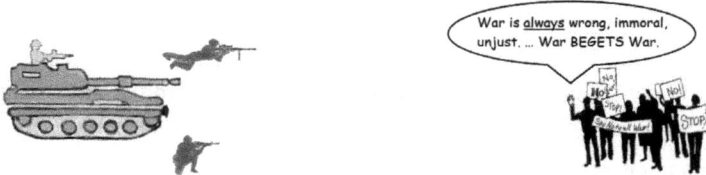

War is <u>always</u> wrong, immoral, unjust. ... War BEGETS War.

• Through legislation, regulations, protest, lobbying, and a dramatic change in buying habits, force corporations and other primary manufacturers, processors, and packaging to appropriately deal through holistic life cycle-, embodied human appropriated net primary productivity- and emergy- analysis, etc., with all negative externalities like pollution, trash, waste, loss of biodiversity, and resistance to antibiotics.

Life cycle analysis

Sure, we can work together.

Farmers Market

Buy Local

When we begin to propose the aforementioned ethos and actions to others, we almost immediately receive "a deer in the headlights," "You are crazy!," or "Yeeaaah. Right." look/response. These general responses come from close family and friends, "Progressives," "Nazis," Christians, Muslims, ignostics, capitalists, socialists, anarchists, etc., etc., i.e., almost everyone.

Herein, we are making another stab at articulating what is inevitably, logically, and ecologically, morally, and ethically, necessary! We are hoping it might open constructive and robust dialogue in community toward a goal of quality life for all.

....

Some years ago the non-profit organization, Ogallala Commons, had its first community intern, Angela Ludolph, in the Seguin area, and Angela wrote the following in her portfolio for this internship (something which is very relevant to the mission of this little book on applied ecology):

"Sustainable agriculture, which should always be a community affair and involve community gardens, was addressed via the U.S. 1985 Farm Bill as LISA, Low Input Sustainable Agriculture. That's the way a conserving, resilient, sustainable agriculture has to be: An agriculture which puts the Precautionary Principle up front and is slow with respect to energy flux and transformation, and material flow, and that places women and a mothering instinct in the driver's seat. Nevertheless, barriers to what is truly sustainable agriculture -and low input, and community, and gardening -were quickly set in place in the 1980s via controlling high-energy (fossil energy and fossil energy-dependent high-input 'renewables'), status quo interests."

For the most part, the authors of this little applied ecology book do not think many of the currently in-vogue "sustainable" practices (or ones proposed to be "the answer"), such as biochar or application of sea weed solutions or high input systems of photovoltaics, electric or smart-hybrid cars, hydroponics, conventional organic food production, or other high-input/-throughput systems are moving us much toward ecological community resilience and sustainability. On the other hand what makes more sense in terms of quality life for all are planned controlled rotational grazing/browsing for appropriately utilizing native grassland/savanna systems (and even for utilizing and managing vegetation within forests and urban areas); local low-input production of appropriate types of vegetables, grains, herbs, and fruits, and domesticated animals as well as honey production; passive solar designs of buildings; lowering consumption by individuals and populations; low-input and -throughput/appropriate-technology; and the development of small schools, holistically integrated with the Land/Nature and with ecology across the curricula and campuses.

Moreover, if any of us are truly going to be of help as Elders -or Masters of anything, we must be first be generalists/master naturalists/positively ethical applied community ecologists. Our goal and process of conservation and development of sustainable ecological community should be primarily reliant on inputs from the local community and not dependent on grants, fossil/mined energy, and materials from outside (especially including plastics). As much as possible, the systems we develop as wise ones should be relatively closed regarding energy flux and material flow and should be mostly utilizing local solar energy received daily.

What follows are several major reasons that development of low input regeneration and conservation of resilient, sustainable ecological community is largely ignored in South Central Texas and other parts of the world:

1. Our socio-political/economic systems emphasize an artificially built environment, mechanization, and instant gratification, resulting in destroying Nature and natural cycles and processes. They reward quantity over quality which results in an increasing accumulation of unnecessary material goods which are not conducive to healthy life for all (including other species). They do not reward ecological soundness and resilience. Moreover, they do not adequately reward social justice and humaneness.

2. Our education systems do a very poor job in facilitating a development of knowledge of ecological principles and processes.

3. We do not critically think about:

• why Nature and the natural and the sense of ecological community and place are necessary,

• the fact that a high rate of local and global energy transformation by humans is harmful to quality life,

• what quality life means as individuals and in community (locally and globally, and including other species), and

- how we will effect effective change toward sustainable community.

Other barriers to a process of sustainability are discussed in blog posts at *paulpeaceparables.com* and *bannedbookscafeblogspot.com* .

Therefore, for these aforementioned reasons we do have much difficulty in sitting down at a common table and in beginning to effectively communicate about what quality life means and how we might go about realizing it, i.e., what our local and global community goals are and what are the objectives, action items, and assessment tools for realizing these.

Even though we the collaborators on this book do recognize that we have not personally done much of significance in our many years of life and that we are all sinners and know we can do better, today is the proverbial first day of the rest of our lives. Therefore, we can do better, and we can begin now. There is hope! Moreover, we strongly feel that efforts toward smaller schools and ecology across curricula/campuses is where all of us should be expending much of our energy.

We know that for those who are broadly and deeply learned, experienced, moral, and ethical, most of what is in this little book on applied ecology is obvious and is nothing new! In addition, we know that for these wise Elders as well as those of us who are wet behind the ears and less learned and experienced, worldwide peace is the goal.

Finally, collaborator paul martin heard from a Franciscan priest in his youth and was impressed by the following biblical quote: "I know thy works, that thou art neither cold nor hot. I would thou wert cold or hot. But because thou art lukewarm and neither cold nor hot, I will begin to vomit thee out of my mouth." paul and the collaborators on this little book DO NOT want anyone to be comfortable, complacent, entirely peaceful until all in the world have peace.

PEACE, or positively ethical applied community ecology, is an agitation process toward peace.

This might be the most important section! *Por favor,* **study it!**

Glossary

(Terms as used in or associated with this little book.
This is placed upfront in hopes of facilitating communication
with the reader. Communication is essential but generally
frustratingly difficult.)

Agroecology (hopefully **"appropriately applied agroecology."** An ecological approach to agriculture that views agricultural areas as ecosystems and is concerned with the ecological impact of agricultural practices. Merriam-Webster

[Philosopher Timothy Morton proposes and ponders through the term and concept of **agrilogistics**, i.e., "agriculture that arose in the Fertile Crescent and that is still plowing ahead. Logistics, because it is a technical, planned, and perfectly logical approach to built space. Logistics, because it proceeds without stepping back and rethinking the logic. A viral logistics, eventually requiring steam engines and industry to feed its proliferation." *http://www.changingweathers.net/en/episodes/48/what-is-dark-ecology]*

Appropriate (often used herein) — "suitable or proper in the circumstances." Google dictionary "Appropriate technologies or practices;" "appropriate goals, policy, plans, actions, monitoring, assessment;" etc. may be very inappropriate if overdone and inappropriate in particular circumstances, spaces, "wholes", times, or geo-political situations.

Artificialization — Process of artificializing. *https://en.wiktionary.org/wiki/artificialization* - Artificial herein is referring to human-dominated systems which have largely lost touch with the important components of natural systems, healthy living soils and waters, photosynthesis and net primary productivity, high biodiversity, and sustainable ecological community dynamics. (Much of today's economy is very artificial and superficial, including conventional and "organic" agriculture, and is not in tune with natural biogeochemical cycles and energetics, and relatively stable local community social fabrics which often existed pre-agriculture, pre-industrialization, or even pre-WW II and the information age.)

Autotrophs & Heterotrophs-Autotrophs — "producing" organisms capable of making nutritive organic molecules from inorganic sources via photosynthesis (involving light energy) or chemosynthesis (involving chemical energy). **Heterotrophs** are "consuming organisms which feed on organic matter produced by, or available in, other organisms." *http://www.biology-online.org*

Biocapacity — Capability of an area to generate an on-going supply of renewable resources and absorb spillover wastes.

Biodiversity — "the variability among living organisms from all sources including, inter [& intra] terrestrial, marine, and other aquatic ecosystems and the ecological complexes of which they are part; this includes diversity within species, between species and of ecosystems." "Biodiversity is the foundation of ecosystem services to which human well-being is intimately linked." *www.greenfacts.org https://www.economist.com/science-and-technology/2018/05/26/a-planetary-census-puts-humans-in-their-place*

Biogeochemical cycles — pathways "by which a chemical substance moves through both the biotic (biosphere) and abiotic (lithosphere, atmosphere, and hydrosphere) components of Earth." "The circulation of chemical nutrients like carbon, oxygen, nitrogen, phosphorus, calcium, and water, etc. through the biological and physical world." *https://en.wikipedia.org*

Biophilia — "coined by Erich Fromm in *The Heart of Man: Its Genius for Good and Evil (1964)* to mean 'love for humanity and nature, and independence and freedom;' extended by Edward O. Wilson in Biophilia (1984) to mean 'the rich, natural pleasure that comes from being surrounded by living organisms'." *Dictionary.com*

Carrying capacity — Number of people (and their consumptive habits) and/or other organisms which a region can support w/out environmental degradation.

Cell — "a small, usually microscopic mass of protoplasm bounded externally by a semipermeable membrane, usually including one or more nuclei and various other organelles with their products, capable alone or interacting with other cells of performing all the fundamental functions of life, and forming the smallest structural unit of living matter capable of functioning independently" Merriam-Webster

- **Prokaryote** — "a small usually microscopic mass of protoplasm bounded externally by a semipermeable membrane, usually including one or more nuclei and various other organelles with their products, capable alone or interacting with other cells of performing all the fundamental functions of life, and forming the smallest structural unit of living matter capable of functioning independently." *https://www.britannica.com/science/prokaryote* "The oldest known fossilized prokaryotes were laid down approximately 3.5 billion years ago, only about 1 billion years after the formation of the Earth's crust. Eukaryotes only appear in the fossil record later, and may have formed from endosymbiosis of multiple prokaryote ancestors." *https://en.wikipedia.org/wiki/Prokaryote*

• **Eukaryote** — "any cell or organism that possesses a clearly defined nucleus. The eukaryotic cell has a nuclear membrane that surrounds the nucleus, in which the well-defined chromosomes (bodies containing the hereditary material) are located. Eukaryotic cells also contain organelles, including mitochondria (cellular energy exchangers), [ribosomes], a Golgi apparatus (secretory device), an endoplasmic reticulum (a canal-like system of membranes within the cell), and lysosomes (digestive apparatus within many cell types). There are several exceptions to this, however, for example, the absence of mitochondria and a nucleus in red blood cells and the lack of mitochondria in the oxymonad *Monocercomonoides* species. Eukaryotes are thought to have evolved between about 1.7 billion and 1.9 billion years ago. The earliest known microfossils resembling eukaryotic organisms date to approximately 1.8 billion years ago." *https://www.britannica.com/science/eukaryote*

• **Organelle** — "The term organelle is derived from the word 'organ' and refers to compartments within the cell that perform a specific function. These compartments are usually isolated from the rest of the cytoplasm through intracellular membranes. These membranes could be similar to the plasma membrane or made from a different complement of lipids and proteins. The properties of a membrane are due to its origin, such as with mitochondria [with its own DNA] or plastids, or due to its specific function, as seen with the nuclear membrane. A few organelles are not membrane-bound and are present as large complexes made of RNA and protein, such as ribosomes." *https://biologydictionary.net/organelle/* (Study the endosymbiosis theory!)

• **Rickettsia** — "any of a various gram-negative, parasitic bacteria [prokaryotes] … that are transmitted by biting arthropods (such as lice or ticks) and cause a number of serious diseases (such as Rocky Mountain spotted fever and typhus)" *https://www.merriam-webster.com/dictionary/rickettsia*

• **Virus** — "an infective agent that typically consists of a nucleic acid molecule in a protein coat, is too small to be seen by light microscopy, and is able to multiply only within the living cells of a host." Google Dictionary

- **Prion** — "an abnormal form of a normally harmless protein found in the brain that is responsible for a variety of fatal neurodegenerative diseases of animals, including humans, called transmissible spongiform encephalopathies." Encyclopedia Britannica

Consilience — "agreement between the approaches to a topic of different academic subjects, especially science and the humanities." Oxford Dictionary [See *https://archive.nytimes.com/www.nytimes.com/books/98/04/26/reviews/980426.26kelvest.html* for a NYT review of E.O. Wilson's book, Consilience.]

Conspicuous consumption — Transformation of energy and consumption of goods at a relatively lavish, glamorous (and very unethical & nonsustainable) scale.

Critical thinking and decision-making — involves thorough and objective analysis of an issue or situation using related facts, evidence, data. Asking the right questions and identifying the problem is of the utmost importance. This and other steps necessitate effective research and recognizing biases. Then one must logically infer and apply or make an educated guess toward appropriate decisions which consider implications and unintended consequences. Without profound holistic knowledge of ecological principles, processes and values, one cannot make educated guesses and critically think or decision-make.

Curricula — "totality of student experiences that occur in the educational process." *https://en.wikipedia.org* Herein this book we use curricula for experiences in any continuing education process, including in governmental entities, non-governmental not-for-profits, businesses, etc. We hope for PEACE, or Positively Ethical Applied Community Ecology across curricula.

Death ideology — An ideology through which we are destroying our own humanity and killing Earth because of a fallacy of misplaced concreteness which allows us "to measure our success in term of gross national product without asking what is being produced, to whom it is being distributed, or what the production process is doing to the worker or the environment." L.E. Schmidt, L.E. & S. Merrato. 2008. The End of Ethics in a Technological Society.

Deep ecology — "environmental philosophy and social movement based in the belief that humans must radically change their relationship to nature from one that values nature solely for its usefulness to human beings to one that recognizes that nature has an inherent value. Sometimes called an "ecosophy," deep ecology offers a definition of the self that differs from traditional notions and is a social movement that sometimes has religious and mystical undertones." Encyclopedia Britannica

Earth —

• **Commons** — The Commons "is a general term referring to the cultural and natural resources accessible to all members of a society, including natural materials such as air, water, and a habitable earth. These resources are held in common, not owned privately." *http:// en.wikipedia.org/wiki/Commons*

• **Land** — The Land is Nature with humans in it who have significant knowledgeable interactions with and consideration of non-human elements such as soils, waters, plants, animals, and other biota. Aldo Leopold can largely be credited with a "Land Ethic".

• **Nature-Living systems** on Earth which are somewhat as they were 15,000 years ago. (I, paul bain martin, have begun using the term dynamic homeostatic symbioses instead of Nature, in trying to sort of deal with philosopher Timothy Morton's logic concerning Nature. *lab.cccb.org/en/tim-morton-ecology-without-nature* "Nature is a sort of anthopocentrically scaled concept, designed for humans, so it's not strictly relevant to thinking about ecology." Timothy Morton)

Eaarth — the new Earth of the Anthropocene, the period during which human activity has been the dominant influence in producing an environment detrimental to life as humans have known it, perhaps beginning in ca. 1950, but maybe much earlier. From Bill McKibben, Eaarth, 2010. (Think of the extra "a" as emphasizing artificial, agrologistics and Anthropocene.)

Ecosystem — abiota and biota interacting in a relatively closed system in an area of the ecosphere/Earth/Eaarth/world. **Ecological Community** -all of the organisms interacting in an area. **Population** -the individuals of one species interacting in an area. **Deme** -a very local sector of a population.

Ecological footprints — "the impact of human activities measured in terms of the area of biologically productive land and water required to produce the goods consumed and to assimilate the wastes generated."

"It is the amount of the environment necessary to produce the goods and services necessary to support a particular lifestyle." *http://wwf.panda.org*

Elder — "someone who has gained recognition as a custodian of knowledge and lore, and who has permission to disclose knowledge and beliefs." *http://www.indigenousteaching.com*

Energy, free energy, energy transformation, energetics, & the energy pyramid — Energy is the ability to move and build things or the capacity to do work. **Free energy**, or Gibbs free energy, is the amount of usable energy or energy that can do work in a system. **Energy transformation**, or a changing to various forms, may involve electrical, thermal, nuclear, mechanical, electromagnetic, sound, and chemical forms. As it is transformed it is degraded, i.e., entropy (see definition below, 2nd Law of TD) is increased. **Energetics** deals with the properties of energy and the way in which it is redistributed in physical, chemical, or biological processes. The energy pyramid in biology illustrates how useful energy (ca. 90%) is lost to entropy as it moves from the autotrophic-producer trophic or food chain level to heterotrophic levels.

Energy units —

- **British thermal unit** — "amount of heat required to raise the temperature of one pound of water by one degree Fahrenheit." (1055 joules or 252 calories) Google dictionary

- **Calorie** — "the energy needed to raise the temperature of 1 gram of water through 1 °C (or 4.1868 joules)." Google dictionary (A kilocalorie is the large Calorie that we commonly use in the U.S.A. for measuring energy; 1000 calories equals a Calorie or a kilocalorie.)

- **Joule** — "one joule is equal to the work done by a force of one newton when its point of application moves one meter in the direction of action of the force, equivalent to one 3600th of a watt-hour." Google dictionary

Equity — fair and just such that everyone is a success vs. equality in which everyone is treated the same.

EROEI — "In physics, energy economics, and ecological energetics, energy returned on energy invested (EROEI or ERoEI); or energy return on investment (EROI), is the ratio of the amount of usable energy (the exergy) delivered from a particular energy resource to the amount of exergy used to obtain that energy resource." *https://en.wikipedia.org* Related pieces: *https://www.scientificamerican.com/article/eroi-behind-numbers-energy-return-investment/ https://ourfiniteworld.com/2016/12/21/eroei-calculations-for-solar-pv-are-misleading/ https://en.wikipedia.org/wiki/Energy_returned_on_energy_invested*

Ethics, Ethos; Morals, Mores; Values —

• **Values** — "The basis of harmony in relationships. Values are intrinsic principles that govern relationships. If one lives in accordance with values in relationships then those relationships become balanced and both the related persons feel the joy in that relationship. For example, if one lives with mutual respect, trust, affection, gratitude then that relationship becomes balanced and harmonized. In this one can see that values are the basis for harmony in relationships. Valueless living is unfulfilled life full of mistakes."

• **Morals** — "The basis of harmony in community. Morals are intrinsic principles that govern community living. Morals are mainly in terms of 'earning wealth', 'marital propriety' and 'propensity towards kindness/cruelty in work-behavior'. If one generates wealth through our own genuine efforts without cheating or stealing, then it is considered righteous wealth which is considered a high moral value. Secondly, if one maintains marital propriety by being sincere, loyal & committed, then we have a righteous spouse of a high moral value. Thirdly, if a person in his-her daily interactions exhibits propensity towards kindness/nurturing in work-behavior instead of cruelty/exploitation, then that is considered high moral value. Immoral living is sin and leads to crime."

• **Mores** – "are customs and conventions in a culture dictated by a society's values."

• **Ethics** — "The basis of harmony in society & social order. Ethics is the policy of living in the society. It is a way of living which nurtures the order in society. The order in society needs to be established and sustained to ensure the continuity of the humankind from one generation to the next forever. This can only be achieved if the resources in form of mind, body and material are purposefully utilized and protected from generation to generation. The policy to do the same is by abundant production, proper distribution and full utilization of all the resources for social welfare so that there is no scarcity, no waste, no deprivation, no exploitation. The sustainability of these resources can be achieved by 1) Proper education, 2) Guarding natural cycles, 3) Protecting the sources of resources, 4) Protecting the goods in transit and storage etc. Only such practices can ensure the continuity of availability of the splendor of this life sustaining planet for the future generations. To live in accordance with these principles is to be ethical. To waste, to horde, to deprive, to destroy, is to be unethical. Unethical living leads to imbalance, exploitation, struggle and conflict/war." *https://www.quora.com/What-is-the-relationship-between-ethics-values-morals-and-attitude*

- **Ethos** — is guiding beliefs or ideals of a society.

Ethic of reciprocity/The Golden Rule — that we strive for equity for all (including other species). All humans should enjoy basic human rights, including access to potable water, food, health care, etc. as rights. For some philosophic reflections on the Golden Rule and ecological ethics, see: *https://philosophynow.org/issues/125/The_Golden_Rule_Revisited https://nome.unak.is/wordpress/08-3/c69-conference-paper/responsibility-to-nature-hans-jonas-and-environmental-ethics/*

Evapotranspiration rate — "Evapotranspiration is the water loss occurring from the processes of evaporation and transpiration. Evaporation occurs when water changes to vapor on either soil or plant surfaces. Transpiration refers to the water lost through the leaves of plants." Evapotranspiration rates are a major factor in determining biomes/ecosystem/ecological communities of a region. *http://ccc.atmos.colostate.edu*

Externalities — usually a cost but a "cost or benefit that affects a party who did not choose to incur that cost or benefit." Ecologists generally strive for policies that internalize an externality, so that costs and benefits will affect mainly parties who choose to incur them. *https://en.wikipedia.org/wiki/Externality*

Food web (or "special" species therein this interconnection of food chains) —

- **Climax** — "species that will remain essentially unchanged in terms of species composition for as long as a site remains undisturbed." *https://en.wikipedia.org*

- **Decreasers** — species which become scarce under herbivory

- **Dominant** — "one of a small number of species which dominate in an ecological community" *https://en.wikipedia.org*

- **Increasers** — species favored under herbivory

- **Keystone** — "a species on which other species in an ecosystem largely depend, such that if it were removed, the ecosystem would change drastically." Google dictionary

"Forty-five/45" — Donald John Trump. He and his tumultuous and turbulent Trumpian administration's policies and actions (e.g., those of Scott Pruitt, Ryan Zinke, Jeff Sessions, Mick Mulvaney, Elisabeth DeVos, Rick Perry, and others) enabled by his Trumpsters, have been the antithesis of being *sabido*/wise, simple, small, slow, steadfast, sharing, and sustainable. (Of course, this business of not practicing good ecological ethics is not new to humankind. Moreover, we all have some Trump in us, and we all had a part in deplorably enabling Trump.)

Fossil energy — Fossil energy sources, including coal, oil, shale, tar sands, and natural gas, formed when prehistoric plants and animals died and their chemical bonds (energy) were gradually buried/protected from decomposition by layers of rock. (Fossil energy has been a major driver in the Anthropocene for agrilogistics and major generator for the hyperobjects of artificial CO_2 and global warming.)

Genetics — "the science of heredity dealing with resemblances and differences of related organisms resulting from the interaction of their genes and the environment." *Dictionary.com* **Epigenetics** — is the study of changes in gene function that are heritable but that are not attributed to alterations of the DNA sequence [DNA methylation and histone modification]. *National Human Genome Res. Instit., NIH*

Geothermal energy — "heat from the Earth. It's [relatively] clean and sustainable. Resources of geothermal energy range from the shallow ground to hot water and hot rock found a few miles beneath the Earth's surface, and down even deeper to the extremely high temperatures of molten rock called magma." *http://www.renewableenergyworld.com*

Haber-Bosch process — "a complex chemical procedure that takes nitrogen from the air [78% nitrogen] and under high pressures and temperatures combines it with hydrogen to produce ammonia. This ammonia is the base of the synthetic nitrogen fertilizers increasingly used around the world today. Nitrogen, a key component of all proteins, DNA, and RNA, is vital to life here. Plants can only use fixed nitrogen and the lack of fixed nitrogen is often the limiting factor in an ecosystem or for crops. Our ability to fix nitrogen ourselves seemed to be a perfect solution. However, the enormous blessings of the Haber Process are balanced by some serious curses.

On the blessing side synthetic nitrogen fertilizer produced by the Haber Process is credited with feeding a third to half the present world population. In fact, about half the nitrogen in each of our bodies is there thanks to the Haber Process. On the cursed side we have several issues including:

• Serious imbalances to the nitrogen cycle.

• High fossil fuel energy inputs.

• Negative effects on soil organisms and soil organic matter.

• Excess runoff causes ocean dead zones.

• Production of a major component of weapons including all those roadside bombs." *www.the-compost-gardener.com*

Haves and have-nots — "the people who are very [or relatively] wealthy and the people who are very poor." *https://www.collinsdictionary. com*

Holistic Management or Holistic Resource Management — "a systems thinking approach to managing resources" *https://en.wikipedia.org/ wiki/Holistic_management_(agriculture)* paul bain martin believes HRMer/ HMer, Allan Savory, did very significantly and positively contribute to pragmatic systems thinking with his little "decision-making framework" or thought-model. *https://www.context.org/iclib/ic25/wood/* However, he did go somewhat overboard in promoting his holistically-planned/controlled rotational-grazing ideas as THE solution to many of the world's ecological problems (especially when he did not always have a good scientific foundation for what he was pontificating).

Homeostasis — The tendency of an organism or a cell (organ system, individual, deme, population, ecological community, ecosystem, Nature, all of the living Earth) to regulate its internal conditions, usually by a system of feedback controls, so as to stabilize health and functioning, regardless of the outside changing conditions. *http://www.biolo-*

gy-online.org There are limits to homeostasis. The Gaia hypothesis, also known as the Gaia theory or the Gaia principle, proposes that organisms interact with their inorganic surroundings on Earth to form a synergistic self-regulating, complex system that helps to maintain and perpetuate the conditions for life on the planet. Topics of interest include how the biosphere and the evolution of life forms affect the stability of global temperature, ocean salinity, oxygen in the atmosphere, the maintenance of a hydrosphere of liquid water and other environmental variables that affect the habitability of Earth.

The Gaia hypothesis was formulated by the chemist James Lovelock and co-developed by the microbiologist Lynn Margulis [who married Carl Sagan] in the 1970s. *https://en.wikipedia.org*

Hubris — Excessive pride or self-confidence [e.g., 45] Google dictionary. Extreme hubris (our inherent ignorance plus a severe lack of humility) enabled humanity to become an agent of global destruction.

Hyperobjects — "Objects [states of matter/energy] that are so massively distributed in time and space as to transcend spatiotemporal specificity, such as global warming, styrofoam, and radioactive plutonium." *https://en.wikipedia.org/wiki/Timothy_Morton*

Ignorance — Lack of knowledge or information. Google dictionary This is the state of humans, even collectively. We have much to learn, much knowledge to gain. (Yet we continue to reproduce and rampantly consume and develop with hubris as if we know it all. We must begin to abide by the Precautionary Principle.)

Isms —

- **Anarchism** — "belief in the abolition of all government and the organization of society on a voluntary, cooperative basis without recourse to force or compulsion." Google dictionary "Anarchists disdain the customary use of 'anarchy' to mean 'chaos' or 'complete disorder'. For them it signifies the absence of a ruler or rulers, a self-managed society, usually resembling the co-operative commonwealth that most socialists have traditionally sought, and more highly organized than the disorganization and chaos of the present. An anarchist society would be more ordered because the political theory of anarchism advocates organization from the bottom up with the federation of the self-governed entities as opposed to order being imposed from the top down upon resisting individuals or groups." *https://www.theguardian.com*

- **Conservativism** — "political doctrine that emphasizes the value of traditional institutions and practices." *https://www.britannica.com* (However, there should be some adherence to a deeper meaning of the root of the word, to the valuing and conserving of Nature and the natural resource base, and to developing a sustainable human social fabric.)

- **Conventional capitalism** — "an economic system characterized by private or corporate ownership of capital goods, by investments that are determined by private decision, and by prices, production, and the distribution of goods that are determined mainly by competition in a free market." Merriam-Webster … The "idea of self-interest is the foundation of conventional capitalism. Conventional capitalism … self-interest, competition, deregulated markets that naturally adjust themselves, resource exploitation, … profit above all else. Enter the age of environmentalism. Suddenly the self-interested devourers of resources realized that the world was not as big as they originally thought. Those resources were not infinite. And exploiting those resources for profit was making our world uninhabitable. Air, water and land quality were being polluted with hazardous and poisonous chemicals. There was a massive hole in the ozone layer. The excess carbon in the atmosphere was causing global warming." *https://advanceconsultingforeducation.wordpress.com*

- **Egalitarianism** — A belief in human equality especially with respect to social, political, and economic affairs. Merriam-Webster

- **Fascism** — A political philosophy, movement, or regime that exalts nation and often race above the individual and that stands for a centralized autocratic government headed by a dictatorial leader, severe economic and social regimentation, and forcible suppression of opposition. Merriam-Webster

- **Idealism** — "The tendency to represent things in an ideal form, or as they might or should be, rather than as they are, with an emphasis on values." Merriam-Webster

- **Liberalism** — "a broad spectrum of political philosophies that consider individual liberty to be the most important political goal, and emphasize individual rights and equality of opportunity. Although most liberals would claim that a government is necessary to protect rights; different forms of liberalism may propose very different policies." *http://www.philosophybasics.com*

- **Libertarianism** — "An extreme laissez-faire political philosophy advocating only minimal state intervention in the lives of citizens." Google dictionary (Compare to anarchy.)

- **Neoliberalism** — Neoliberalism sees competition as the defining characteristic of human relations. It redefines citizens as consumers, whose democratic choices are best exercised by buying and selling, a process that rewards merit and punishes inefficiency. It maintains that "the market" delivers benefits that could never be achieved by planning.

"Attempts to limit competition are treated as inimical to liberty. Tax and regulation should be minimized, public services should be privatized. The organization of labor and collective bargaining by trade unions are portrayed as market distortions that impede the formation of a natural hierarchy of winners and losers. Inequality is recast as virtuous: a reward for utility and a generator of wealth, which trickles down to enrich everyone. Efforts to create a more equal society are both counterproductive and morally corrosive. The market ensures that everyone gets what they deserve." *https://www.theguardian.com/books/2016/apr/15/neoliberalism-ideology-problem-george-monbiot*

- **Oligarchism** — The principle or spirit of an oligarchy in which "a small group of people have control of a country, organization, or institution." Google dictionary

- **Plutocracism** — The principle or spirit of a plutocracy in which you have "an elite or ruling class of people whose power derives from their wealth." Google dictionary

- **Postmodernism** — "Postmodernism is largely a reaction to the assumed certainty of scientific, or objective, efforts to explain reality. In essence, it stems from a recognition that reality is not simply mirrored in human understanding of it, but rather, is constructed as the mind tries to understand its own particular and personal reality. For this reason, postmodernism is highly skeptical of explanations which claim to be valid for all groups, cultures, traditions, or races, and instead focuses on the relative truths of each person. In the postmodern understanding, interpretation is everything; reality only comes into being through our interpretations of what the world means to us individually. Postmodernism relies on concrete experience over abstract principles, knowing always that the outcome of one's own experience will necessarily be fallible and relative, rather than certain and universal. Postmodernism is 'post' because it denies the existence of any ultimate principles, and it lacks the optimism of there being a scientific, philosophical, or religious truth which will explain everything for everybody - a characteristic of the so-called 'modern' mind. The paradox of the postmodern position is that, in placing all principles under the scrutiny of its skepticism, it must realize that even its own principles are not beyond questioning." *https://www.pbs.org*

• **Pragmatism** — A philosophy in which "… truth is preeminently to be tested by the practical consequences of belief." Merriam-Webster

• **Progressivism** — Progressivism is the support for or advocacy of social reform. As a philosophy, it is based on the idea of progress, which asserts that advancements in science, technology, economic development, and social organization are vital to the improvement of the human condition. *https://en.wikipedia.org*

According to Senator Elizabeth Warren- "We [Progressives] believe that Wall Street needs stronger rules and tougher enforcement, and we're willing to fight for it." - "We believe in science, and that means that we have a responsibility to protect this Earth." - "We believe that the Internet shouldn't be rigged to benefit big corporations, and that means real net neutrality." - "We believe that no one should work full-time and still live in poverty, and that means raising the minimum wage." - "We believe that fast-food workers deserve a livable wage, and that means that when they take to the picket line, we are proud to fight alongside them." - "We believe that students are entitled to get an education without being crushed by debt." - "We believe that after a lifetime of work, people are entitled to retire with dignity, and that means protecting Social Security, Medicare, and pensions." - "We believe — I can't believe I have to say this in 2014 — we believe in equal pay for equal work." - "We believe that equal means equal, and that's true in marriage, it's true in the workplace, it's true in all of America." - "We believe that immigration has made this country strong and vibrant, and that means reform." - "And we believe that corporations are not people, that women have a right to their bodies. We will overturn Hobby Lobby and we will fight for it. We will fight for it!" And the main tenet of conservative philosophy, according to Warren? "I got mine. The rest of you are on your own." *https://www.theatlantic.com*

• **Republicanism** — A desired system of governance: "In a republic, a constitution or charter of rights protects certain inalienable rights that cannot be taken away by the government, even if it has been elected by a majority of voters. In a 'pure democracy,' the majority is not restrained in this way and can impose its will on the minority." *www.diffen.com*

• **Scientism** — the cosmetic application of science in unwarranted situations not covered by the scientific method. *https://en.wikipedia.org*

• **Socialism** — any of various economic and political theories advocating collective or governmental ownership and administration of the means of production and distribution of goods. Merriam-Webster Of course there are varying degrees of socialism —or capitalism, communism, etc.—and they might be preferably democratic or autocratic, etc. **Communism** — "… the political belief that all people are equal, that there should be no private ownership and that workers should control the means of producing things." *https://www.collinsdictionary.com*

• **Sociocracism** — a whole systems approach to designing and leading organizations. It is based on principles, methods, and a structure that creates a resilient and coherent system. It uses transparency, inclusiveness, and accountability to increase harmony, effectiveness, and productivity. *http://thesociocracygroup.com*

• **Tea Party-ism** — a "movement within the Republican Party. Members of the movement have called for a reduction of the U.S. national debt and federal budget deficit by reducing government spending, and for lower taxes. The movement opposes government-sponsored universal healthcare and has been described as a mixture of libertarian, populist, and conservative activism."

• **Totalitarianism** — "absolute control by the state or a governing branch of a highly centralized institution." *www.dictionary.com*

- **Transcendentalism** — "idealistic system of thought based on a belief in the essential unity of all creation, the innate goodness of humanity, and the supremacy of insight over logic and experience for the revelation of the deepest truths." *https://www.britannica.com*

Life-cycle analysis — comprehensive ecological assessment that identifies the energy, material, and waste flows of a product, and their impact on the environment. This cradle to grave evaluation begins with the design of the product and progresses through the extraction and use of its raw materials, manufacturing or processing with associated waste stream, storage, distribution, use, and its disposal or recycling. The objective is to identify changes, at every stage of the life cycle, that can lead to environmental benefits and overall cost savings. *www.businessdictionary.*

Life — Matter organized with inputs of (solar) energy. Ecosystems are important units in the study of life or biology. Major ecosystem parts include mineral cycles, the water cycle, energy flow, and living organisms.

Maslow's needs — "In his influential paper of 1943, A Theory of Human Motivation, the American psychologist Abraham Maslow proposed that healthy human beings have a certain number of needs, and that these needs are arranged in a hierarchy, with some needs (such as physiological and safety needs) being more primitive or basic than others (such as social and ego needs). Maslow's so-called 'hierarchy of needs' is often presented as a five-level pyramid, with higher needs coming into focus only once lower, more basic needs are met." *https://www.psychologytoday.com/blog*

McMansions — "… a pejorative term for a large 'mass-produced' dwelling, constructed with low-quality materials and craftsmanship, using a mishmash of architectural symbols to invoke connotations of wealth or taste, executed via poorly thought-out exterior and interior design." *https://en.wikipedia.org/wiki/McMansion*

Metabolism — "the complex of physical and chemical processes occurring within a living cell or organism that are necessary for the maintenance of life" or "any basic process of organic functioning or operating." *http://www.dictionary.com*

Native biota — biota which existed "naturally" in a region ca. 500 to 20,000 years ago or more and still do.

Natural resource base — soil/air, water, daily solar energy, and diverse biota of an area, and resultant biogeochemical cycles.

Net primary productivity/human appropriated net primary productivity (HANPP)/embodied HANPP — the rate at which an ecosystem accumulates energy or biomass, excluding the energy it uses for the process of respiration. This typically corresponds to the rate of photosynthesis, minus respiration by the photosynthesizers. *http://www.biology-online.org* HANPP is "an integrated socioecological indicator quantifying effects of human-induced changes in productivity and harvest on ecological biomass flows." Google dictionary. eHANPP "accounts allocate to any product the entire amount of the human appropriation of net primary production (HANPP) that emerges throughout its supply chain. This allows consumption-based accounts to move beyond simple area-demand approaches by taking differences in natural productivity as well as in land-use intensity into account, both across land-use types as well as across world regions." *https://wiki.p2pfoundation.net/Embodied_Human_Appropriation_of_Net_Primary_Production*

Krausmann et al. 2010 Global human appropriation of net primary production doubled in the 20th century. Proceedings of the National Academy of Sciences of the U.S.A.: "This work analyzes trends in HANPP from 1910 to 2005 and finds that although human population has grown fourfold and economic output 17-fold, global HANPP has only doubled. Despite this increase in efficiency, HANPP has still risen from 6.9 Gt of carbon per y in 1910 to 14.8 GtC/y in 2005, i.e., from 13% to 25% of the net primary production of potential vegetation. Biomass harvested per capita and year has slightly declined despite growth in consumption because of a decline in reliance on bioenergy and higher conversion efficiencies of primary biomass to products. The rise in efficiency is overwhelmingly due to increased crop yields, albeit frequently associated with substantial ecological costs, such as fossil energy inputs, soil degradation, and biodiversity loss. If humans can maintain the past trend lines in efficiency gains, we estimate that HANPP might only grow to 27–29% by 2050, but providing large amounts of bioenergy could increase global HANPP to 44%. This result calls for caution in refocusing the energy economy on land-based resources and for strategies that foster the continuation of increases in land-use efficiency without excessively increasing ecological costs of intensification."

Oxidation & reduction — Oxidation is the loss of electrons or an increase in oxidation state by a molecule, atom, or ion. Reduction is the gain of electrons or a decrease in oxidation state by a molecule, atom, or ion. "Plants represent one of the most basic examples of biological oxidation and reduction. The chemical conversion of carbon dioxide and water into sugar (glucose) and oxygen is a light-driven reduction process:

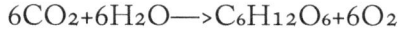

$$6CO_2 + 6H_2O \longrightarrow C_6H_{12}O_6 + 6O_2$$

The process by which non-photosynthetic organisms and cells obtain energy, is through the consumption of the energy rich products of photosynthesis. By oxidizing these products, electrons are passed along to make the products carbon dioxide, and water, in an environmental recycling process. The process of oxidizing glucose and atmospheric oxygen allowed energy to be captured for use by the organism that consumes these products of the plant. The following reaction represents this process:

$$C_6H_{12}O_6 + O_2 \longrightarrow 6CO_2 + 6H_2O + Energy$$

It is therefore through this process that heterotrophs (most generally 'animals' which consume other organisms obtain energy) and autotrophs (plants which are able to produce their own energy) participate in an environmental cycle of exchanging carbon dioxide and water to produce energy containing glucose for organismal oxidation and energy production, and subsequently allowing the regeneration of the byproducts carbon dioxide and water, to begin the cycle again. Therefore, these two groups of organisms have been allowed to diverge interdependently through this natural life cycle." *https://chem.libretexts.org*

PEACE — A relatively tranquil and harmonious state without war and violence in a biotic system. Relative to human policy & actions, it should be a process of "Positively Ethical Applied Community Ecology".

Permaculture — "is a system of agricultural and social design principles centered around simulating or directly utilizing the patterns and features observed in natural ecosystems. The term permaculture was developed and coined by David Holmgren, then a graduate student, and his professor, Bill Mollison, in 1978. The word permaculture originally referred to 'permanent agriculture', but was expanded to stand also for 'permanent culture', as it was understood that social aspects were integral to a truly sustainable system as inspired by Masanobu Fukuoka's natural farming philosophy.

It has many branches that include, but are not limited to, ecological design, ecological engineering, environmental design, and construction. Permaculture also includes integrated water resources management that develops sustainable architecture, and regenerative and self-maintained habitat and agricultural systems modelled from natural ecosystems."

Mollison has said: "Permaculture is a philosophy of working with, rather than against nature; of protracted and thoughtful observation rather than protracted and thoughtless labor; and of looking at plants and animals in all their functions, rather than treating any area as a single product system." *https://en.wikipedia.org/wiki/Permaculture* (paul bain martin was very disappointed in Mollison when he spoke at the Texas Department of Agriculture, Austin in the 1980's, and has not been impressed with "on the Land" examples of "permaculture" he has seen. However, the enthusiasm of permaculturalists, young & old, is very impressive and hopeful!!)

Policy — Policies can be understood as political, managerial, financial, [legal], and administrative mechanisms arranged to reach explicit goals. *https://en.wikipedia.org/wiki/Policy* Policies can be thought of as objectives to reach a goal and guidance for decisions and actions.

Poverty and extreme poverty — "Extreme poverty, absolute poverty, destitution, or penury, was originally defined by the United Nations in 1995 as 'a condition characterized by severe deprivation of basic human needs, including food, safe drinking water, sanitation facilities, health, shelter, education and information. It depends not only on income but also on access to services.' In 2008, 'extreme poverty' widely refers to earning below the international poverty line of $1.25/day (in 2005 prices), set by the World Bank. This measure is the equivalent to earning $1.00 a day in 1996 US prices ... ." *https://en.wikipedia.org/wiki/Extreme_poverty*

Power — can mean control and influence over others, including other species and the natural resource base. It is also the time rate at which energy is transformed. Humans who have influence over energy transformation are very powerful.

Precautionary Principle — "the concept that establishes it is better to avoid or mitigate an action or policy that has the plausible potential, based on scientific analysis, to result in major or irreversible negative consequences to the environment or public even if the consequences of that activity are not conclusively known, with the burden of proof that it is not harmful falling on those proposing the action. It is a major principle of international environmental law and is extended to other areas and jurisdictions as well." *http://www.newworldencyclopedia.org*

Quality life — good health, education, time for play and rest, and plenty of good friends. Healthy individuals, demes, populations, ecological communities, ecosystems; a healthy ecosphere. Social justice, humaneness, ecological sanity. Equity.

r- & K-strategists — The organisms described as *r-strategists* typically live in unstable, unpredictable environments. Here the ability to reproduce rapidly (exponentially) is important. Such organisms have high fecundity and relatively little investment in any one progeny individual; they are typically weak and subject to predation and the vicissitudes of their environment. The "strategic intent" is to flood the habitat with progeny so that, regardless of predation or mortality, at least some of the progeny will survive to reproduce. Organisms that are r-selected have short life spans, are generally small, quick to mature, and waste much energy. Typical examples of r-strategists are salmon, corals, insects, and bacteria.

K-strategists, on the other hand occupy more stable environments. They are larger in size and have longer life expectancies. They are stronger or are better protected and generally are more energy efficient. They produce, during their life spans, fewer progeny but place a greater investment in each. Their reproductive strategy is to grow slowly, live close to the carrying capacity of their habitat and produce a few progenies, each having a high probability of survival. Typical K-selected organisms are elephants and humans. *https://www.cs.montana.edu*

Radioisotopes — "any of several species of the same chemical element with different masses whose nuclei are unstable and dissipate excess energy by spontaneously emitting radiation in the form of alpha, beta, and gamma rays." *https://www.britannica.com/science/radioactive-isotope*

"… Radioisotopes are able to help us treat diseases such as cancer. They also enable doctors to identify the specific areas or parts of the body of patients, so that they know where the problem lies. That's what radioisotopes can do for us. They help us in locating problems in bodies; they help us in treating diseases and sicknesses and is widely used in therapies and medicine. They even kill the bacteria in our food, and are sometimes used in the smoke detectors, but as we know, radioisotopes decay as well.

While radioisotopes have a lot of advantages, they have their disadvantages as well. They are radioactive and can be harmful and kill organisms. If the radioisotopes are given or pointed at a part of the body of a person, which is completely normal, then the radioisotopes would be killing healthy cells, and that would be harmful to the human being. And in addition, we know that isotopes are used in the construction of bombs." *https://duy13.wordpress.com/2011/07/08/radioisotopes-good-or-bad-i-say-good-and-bad/*

Religion — (The important stem-root is ligare, or to link together.) "The belief in and worship of a superhuman controlling power, especially a personal God or gods." Google dictionary Karen Armstrong's writings are good on religious history, including A History of God: The 4,000-Year Quest of Judaism, Christianity, and Islam.

Renewable energy — Renewable energy (sources) "capture their energy from existing flows of energy, from on-going natural processes, such as sunshine, wind, flowing water, biological processes, and geothermal heat flows. The most common definition is that renewable energy is from an energy resource that is replaced rapidly by a natural process such as power generated from the sun or from the wind. Most renewable forms of energy, other than geothermal and tidal power, ultimately come from the Sun." *https://www.sciencedaily.com*

Problems which paul bain martin has with the term, "renewable energy," include:

1. The 1st Law of Thermodynamics states that energy cannot be created or destroyed. Therefore, energy isn't truly "renewable."

2. The major source of energy from the sun is relatively constant and dependable, as long as tried-and-true homeostatic "Gaia" processes are not disrupted. However, with the conspicuous consumption of the Haves and population increases of humans and domesticated species, we have overshot homeostasis.

3. The term renewable energy tends to provide current human populations false hopes that they can simply continue to develop and begin to mass-produce highly artificial systems of electric car & hybrid cars, wind-turbines, photovoltaic & other solar-energy transformers, wave- & tide-energy transformers, fuel cells and other systems using batteries, and that this will result in quality life for all including other species. The Truth is that humans must begin to live the 7 S's! (See the

next definition.) *http://www.paulpeaceparables.com/2018/05/01/renewable-energy-as-the-key-asset-of-commonwealth-in-community-by-paul-bain-martin1/*

Sabio (Spanish for wise), Simple, Small, Slow, Steadfast, Sharing, … SUSTAINABLE! (The 7 S's) — A sustainable world of sustainable livelihoods is one in which we live wisely (*sabiamente* in Spanish or Portuguese), with a light individual & collective ecological footprint and low number of kilocalories/joules/BTUs used per capita/day (simple, small, slow), … and which shares in solidarity toward equity.

Savanna & Taiga — Most of the biomes (tundra, temperate deciduous forests, tropical rainforests, and deserts) are known by students. Savanna is a temperate or "a tropical or subtropical grassland biome (as of eastern Africa or northern South America) containing scattered trees and drought-resistant undergrowth," which generally receives regular fires. Taiga is "a moist subarctic forest biome dominated by conifers (such as spruce and fir) that begins where the tundra ends." Merriam-Webster

Science (and sexual persuasion, gender, ethnicity, race, species) — Science is the quest for knowledge. … We know little, but science and the scientific method have allowed us to develop a considerable body of knowledge.

Despite what genetics, environmental influences, traditions, culture, religion have taught us and brain-washed us to believe, there is a wide variation in sexual urges and behavior and physical, physiological, and psychological gender differences in humans and other species. This is natural and can be normal and contribute to the success of a deme, population, ecological community, ecosystem, or the ecosphere. Variation as to ethnicity and resultant diversity can be very healthy.

There is variation in the human species; however, compared to variation in other species, it is relatively small. Therefore, **race** for humans is probably not a very useful term in general, and use of it causes more problems than utility. We humans are all of one race. [This is not to say that inbreeding and out-breeding have not caused clusters of humans who have certain genetically-related problems, such as such as cystic fibrosis, sickle cell anemia, Marfan syndrome, Huntington's disease, hemochromatosis, increased susceptibility to obesity, diabetes, or alcoholism, and that these clusters should not be identified.]

Second Law of Thermodynamics/Entropy — "The level of disorder in the universe is steadily increasing. Systems tend to move from ordered behavior to more random [or disordered] behavior." "A measure of the level of disorder of a system is entropy." As energy is transformed, it tends toward uselessness. Order in one system creates disorder in another, e.g., an air-conditioned Trumpian high-rise with high flush toilets, electric clothes dryers, abundant glamorous lighting, water in plastic bottles, and Trump steaks is very destructive of Nature or dynamic homeostatic symbioses. *http://physics.bu.edu*

Semiochemicals — "a pheromone or other chemical that conveys a signal from one organism to another so as to modify the behavior of the recipient organism." Google dictionary

Social ecology — "the interrelationship between social and natural systems in the context of sustainable development. Research [at the Institute of Social Ecology, Vienna] is structured into four main fields (i) society's metabolism, (ii) land use change and human intervention in natural systems, (iii) environmental history and cultural evolution, and (iv) transition studies." *https://www.sume.at/iff*

Social metabolism or socioeconomic metabolism — the set of flows of materials and energy that occur between Nature and society, between different societies, and within societies. These human-controlled material and energy flows are a basic feature of all societies, but their magnitude and diversity largely depend on specific cultures, or socio-metabolic regimes. Social or socioeconomic metabolism is also described as "the self-reproduction and evolution of the biophysical structures of human society. It comprises those biophysical transformation processes, distribution processes, and flows, which are controlled by humans for their purposes. The biophysical structures of society ('in use stocks') and socioeconomic metabolism together form the biophysical basis of society." *https://en.wikipedia.org/wiki/Social_metabolism*

Soil — The four generalized blocks of an ecosystem, soil, water, solar energy, and biota are all essential. However, it could be argued that perhaps the most important is soil. An excellent soil has good tilth with perhaps equal components of sand, silt, and clay and ca. 5% organic matter; good soil drainage; large populations of microorganisms; and sufficient, but not excessive, levels of essential nutrients.

Species — the major subdivision of a genus or subgenus, regarded as the basic category of biological classification, composed of related individuals that resemble one another, are able to breed among themselves, but are not able to breed with members of another species. *Dictionary.com*

[Species are/is ever-changing/a dynamic process (micro-evolution to evolution)]

Steady-state economics — A steady state economy is an economy with stable or mildly fluctuating size. The term typically refers to a national economy, but it can also be applied to a local, regional, or global economy. An economy can reach a steady state after a period of growth or after a period of downsizing or degrowth. To be sustainable, a steady state economy may not exceed ecological limits. *http://www.steadystate.org*

Sufficiency vs. efficiency — Even if systems or subsystems become more and more efficient, as we consume more, or as systems grow, we continue to exacerbate socio-political/economic (ecological) problems for quality life, including those for other species.

Sustainable livelihoods — A livelihood is sustainable when it can cope with and recover from stresses and shocks and maintain or enhance its capabilities and assets both now and in the future, while not undermining natural resource bases. *http://www.fao.org*

Sustainability — The ability to continue and to provide quality life. It involves fairness, and protection of the environment. Social justice, humaneness, ecological sanity. "Ensuring life on Earth is an infinite game, the endless expression of generosity on behalf of all." Paul Hawken. 2007. Blessed Unrest.

Transdisciplinary research/work — "research efforts conducted by investigators from different disciplines working jointly to create new conceptual, theoretical, methodological, and translational innovations that integrate and move beyond discipline-specific approaches to address a common problem." **Interdisciplinary research**-"any study or group of studies undertaken by scholars from two or more distinct scientific disciplines. The research is based upon a conceptual model that links or integrates theoretical frameworks from those disciplines, uses study design and methodology that is not limited to any one field, and requires the use of perspectives and skills of the involved disciplines throughout multiple phases of the research process." *https://www.hsph.harvard.edu/trec/about-us/definitions/* **Multidisciplinary research**-bringing disciplines together to talk about issues from each of their perspectives. They may collaborate, but they maintain a separation of their disciplines in that process. When the project is done, those disciplines go back to where they came from to start other projects. Google dictionary

Transnational corporations — Transnational corporations are businesses that operate across international borders, although most of them have their headquarters in the USA, Europe and Japan.

There were about 7000 TNC's operating in 1970, but the charity Christian Aid estimates that this figure has now increased to about 63,000 with about 690,000 subsidiaries which operate in almost every sector of the economy and almost every country in the world today.

The key characteristics of TNC's are the following:

They seek competitive advantages and maximization of profits by constantly searching for the cheapest and most efficient production locations across the world.

They have geographical flexibility; they can shift resources and operations to any location in the world.

A substantial part of their workforce is located in the developing world, but often employed indirectly through subsidiaries.

TNC assets are distributed worldwide rather than focused in one or two countries, for example, 17 of the top 100 TNCs have 90% of their assets in a different country from their head office.

TNC's are economically very wealthy and thus potentially more powerful than many of the world's nation states.

Truth — Herein **Truth** is *"what should be."* Today's **reality** is *"what is."* It is "a transcendent fundamental or spiritual reality." *Merriam-Webster*. Also, Truth is subjective or objective, relative or absolute. "We need scientific truths for deriving practical benefits. We need religious truths (truths from the humanities) for deriving inner peace and satisfaction." *http://www.metanexus.net/essay/scientific-and-religious-truths* "Mythology is the penultimate truth; penultimate because the ultimate cannot be put into words. It is beyond words, beyond images. Mythology pitches the mind beyond that rim, to what can be known but not told." Joseph Campbell (For paul bain martin, fundamental truths are the 7 S's; seeking knowledge, wisdom, prudence; recognizing limits; abiding by the Precautionary Principle and the 2nd Law of TD; and living a holistic, comprehensive, and profound Ethic of Reciprocity.)

War — "a state of armed conflict between societies. It is generally characterized by extreme aggression, destruction, and mortality, using regular or irregular military forces. An absence of war is usually called 'peace.'" "The earliest recorded evidence of war belongs to the Mesolithic cemetery Site 117, which has been determined to be approximately 14,000 years old. About forty-five percent of the skeletons there displayed signs of violent death. Since the rise of the state some 5,000 years ago, military activity has occurred over much of the globe." *https://en.wikipedia.org* The root of many human conflicts has been religious differences. [Herein war is considered to be basically the massive conflict of human against Nature.]

Watershed — An area of land, where all the water that is under the land or drains off the land, goes into the same place. Geologists, naturalists, and ecologists like John Wesley Powell and Eugene Odum have proposed that these earth unities would be logical political entities and best for making local and global policy through positively ethical applied community ecology.

Why?? What? How??? — Questions which I (paul bain martin) hope to prod folk into beginning to answer through the publicizing of this little book on "Positively Ethical Applied Community Ecology"

- **Why??** have values which begin to move us toward collective efforts in solidarity to have a healthy Earth, quality life for all humans, and ample quality habitat for all species? Why?? should we live: *Sabido* — ly, Simply, Smally, Slowly, Steadfastly, Sharingly, Sustainably?, i.e., why should we live the 7 S's!?!

- **What?** actions do we need to take to live these S's?

- **How???** do we realize a commonality of the Why?? with others and How??? do we practice a lifestyle of doing the Whats?

[**VV->^^** — We have been using 2 arrows pointing down with a horizontal arrow connecting them with 2 smaller arrows pointing up, to remind us:

- that amount of consumption per capita by the Haves, and growth of human populations and domesticated species (all Haves relative to wild species) must be reduced, and

- that much of the power of the Haves needs to be transferred to human have-nots and to be used to increase quality habitat for other species.]

Balancing Global Issues
for a Healthy Earth

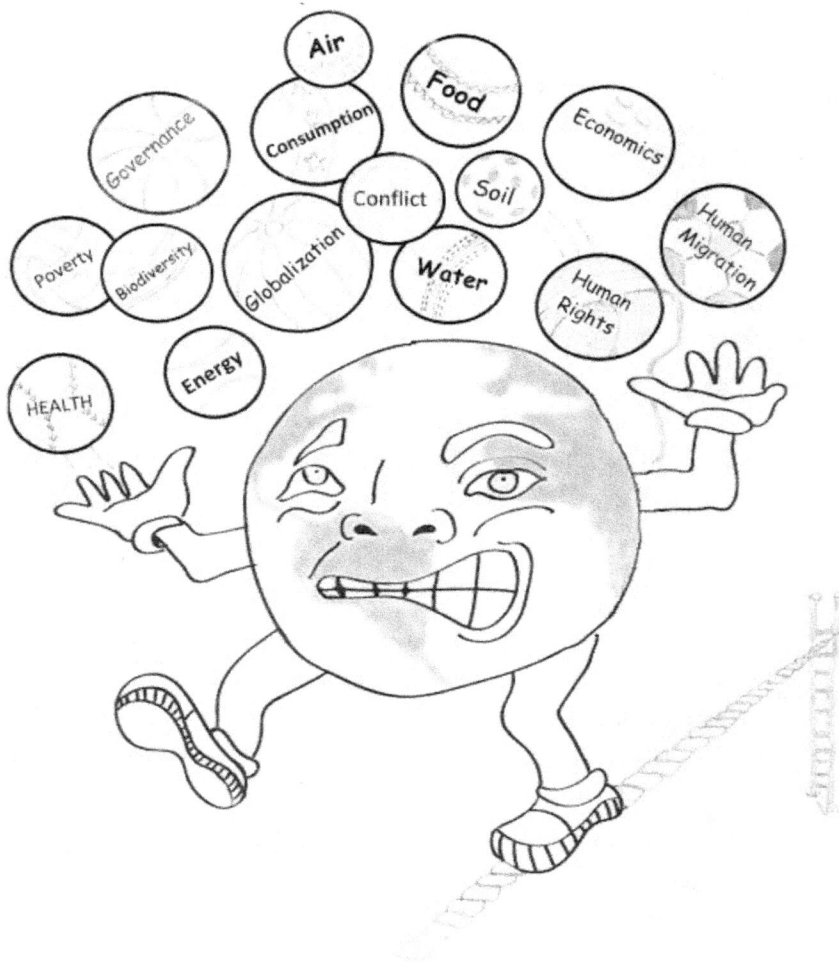

Faith, Hope and Charity within Humankind? Can We Rally Around Limiting embodied Human Appropriated Net Primary Productivity?

Eaarth to Neo-Earth

It is blatantly obvious that
    We must go beyond realpolitik … Or ....
    Realpolitik is now
    What we previously called idealistic.

No! It wasn't Kool-Aid!!

However, I did drink the earthy, healthy *atole* of awareness, hope, and action from the early progressive conservation movement of ca. 1890-1920 and Vatican II, pro-civil rights, anti-War, pro-ecological-sanity sit-ins, marches, songs, poetry, speeches, articles, and books, and effective non-violent protesting (and lobbying and voting) of the 1960's-70's. A bit later, the 1973 crisis in energetics spiked that wonderful elixir. And, more recently the exponential accumulation of challenges of social injustice, inhumaneness, and ecological insanity of the Anthropocene here on this Eaarth has truly strengthened this ancient tonic, which I both metaphorically and actually imbibe routinely.

One very hot afternoon in August when I was about ten years of age and struggling to mow my wealthier Uncle Peggy and Aunt Joe Bailey's expansive lawn possessing steep and very difficult slopes, an extremely beautiful Carmelite nun from Mexico stopped and gave me a glass quart (Ball Mason jar) of cold refreshing *atole*. This young brown-and-white clothed angel was one of several who were teaching us catechism at St. Joseph's Catholic Church that summer. They had been invited for supper into our very small two-bedroom home on the outskirts of Devine, and the chilled *atole* was a gift for my family of eight. ("For my family!" Nevertheless, I quickly drank this unbelievably great manna from heaven all down my young gullet after *la angelica* floated away west on Highway 173 toward town and St. Joseph's on that blistering sunny day.)

I am now in my 70's, my knees are stiff and do not function as they once did. At times I feel really miserable late in the afternoon or when I get out of bed in the morning, and especially after long car trips or working for long periods on *este procesador de textos y máquina de búsqueda de google*. But, I do regularly become refreshed and recharged by *atoles* of very adorable and knowledge-seeking youth with whom I learn about PEACE and kid around—through NGO's like *Kids on the Land, Generations Indigenous Ways on Pine Ridge Reservation, Dos Pueblos (Arco Iris)* in Nicaragua, *Ogallala Commons*, the *Seguin Outdoor Learning Center, HEB* camps on the Frio River, Seguin community gardens and especially the *LULAC* garden, and *Episcopalian* service efforts in Honduras and Mexico as well as in family gatherings with my beautiful grandkids.

A primary reason for this little book of PEACE, is that I want to bring us back from Eaarth to Earth, and to provide for some of the youth of the world some of the dynamic homeostatic symbioses, or "nature," about which I was able to learn and thoroughly experience in my own lifetime. I fervently wish to have a small part in helping to leave a world system of quality life for all, including other species, for as long as possible.

My life experiences and academic learning with respect to realizing sustainable community have primarily been in agriculture and biology, and much of my academic learning has focused on agricultural entomology. Although I can communicate well scientifically to some degree, I do prefer relating personal stories to make my scientific points in a socio-economic/political and psychological (i.e., collectively ecological) context about sustainability. Personal anecdotal situations in which I have been that have a sustainability message of needing to persevere and to live wisely and humbly and in a giving way include the following:

• A plethora of warm tales (as well as those not so warm and fuzzy) from a five-acre diversified hog farm of the Alton and Louise Martin family in a south Texas ecosystem

(Transnational Wal-Mart and other businesses now cover the much of the area which had been mostly photosynthesizing pasture; these largely parasitic structures slickly market stuff which isn't appropriate for realizing sustainable community).

• Almost passing away in an avalanche of cottonseed hulls in a poorly lit, hustle and bustle of a feed mill in my hometown (pre-OSHA).

• Killing three beautiful large male boars my dad had recently bought in the Pearsall area to rapidly finish out and take to the Swift & Co. slaughtering plant in San Antonio; they died because of my rushed castrations in order to make a date with my girl friend in that moment of time;

• Coming close to dying in cotton insect research plots in the Brazos Bottom from the very toxic, cholinesterase-inhibiting biocide, aldicarb, ultimately produced through groupthink and callous transnational corporations (There were other near-death experiences. My mom says I am a cat with nine lives who has used up most of them.);

• Religious and military experiences in the 1960's, and my sort of ignostic-psyche as influenced by this period, my Marine Master Sergeant-dad, and being in a Catholic-Protestant family pre- and post-Vatican II;

• Devastation of non-sustainable monocultures of coastal bermudagrass in the SE U.S. and *Brachiaria* in far western Brasil by fall armyworms and spittlebugs;

• Experiences on the Sugar Land, Texas, prison farms and seeing the brown and black *de facto* slaves work the fields while white trustees on horseback with guns looked over these imprisoned field laborers. (I was there as an entomology student-worker as a result of biocide-resistant tobacco budworms eating up their cotton and subsequent USDA band-aid-type research done for transnational corporate profits.);

• *Las aguas prietas de imperialismo* (Coca-Cola) at the *Colegio Superior de Agricultura Tropical*, Cárdenas, Tabasco, Mexico;

• Land reform and the mosaic of Roberto Strang's fazenda in the Mato Grosso do Sul, Brasil;

• A month of work in a community garden in Krakow, Poland, a garden which changed dramatically after The Wall fell and involvement in other community gardens;

• Gifts of caps, meals, and fishing and hunting trips, and campaign donations mostly from large transnational corporations—power and quid pro quo corruption at Land Grant institutions, experiment stations, and regulatory agencies, including my own failings (Oh come on … you too have experienced some of this in this very capitalistic world!);

• Non-adherence to the Precautionary Principle causing toxins from old tick-control dipping vats within the Alamo Ranch subdivision in San Antonio, and on military bases, along with drenching of nursery products to combat fire ants, chemical spills at the Devine Municipal Airport, etc. to seep into our soils and water;

   • Many other stories, primarily from Texas, which perhaps point out needs for and ways of achieving sustainable community.

These stories deal with human overshoot of carrying capacity, drawdown on the natural resource base, and the tremendous disparity and poverty in the world. They underline the need to live the 7 S's, which means individually and collectively living: *Sabiamente* (Wisely), Simply, Smally, Slowly, Steadfastly, Sharingly, Sustainably. These tales point out the need for a dramatic leveling off of growth of populations of humans and domesticated biota and reduction of consumption by the Haves, as well as a gracious sharing of the power of the Haves with the have-nots of Eaarth (including other species).

Some of these short stories are employed herein this little book or are posted at *www.paulpeaceparables.com/*

Embodied-Human Appropriated Net Primary Productivity. Back in the late 1980's/early 1990's I had some brief and uncomfortable arguments with cantankerous but very intelligent, clairvoyant, and focused Allan Savory and some of his "Holistic Resource Management" disciples about planned-controlled-grazing (which is a good tool for consideration and employment in sustainable systems), Allan's set-in-concrete "rules and taboos," and some of his dogma. Nevertheless, I sincerely and deeply appreciate Allan for helping me think more holistically about what I now mostly call PEACE and the need for developing clear goals and a strategic plan of action. (I have tried to explain some of what I gathered from Allan in several previous papers/publications to which I have contributed.)

Allan emphasizes the importance of monitoring and replanning after actions toward sustainable community. If one does not have sustainable measures of sustainable community and ecosystem assessments, then he/she is mostly living in the dark and is not going to reach profound and holistic quality life for all for a truly significant period. Later in this little book we will present a taxonomy of sustainable indicators which can assist in realizing a goal of sustainability for all, i.e., social justice, humaneness, and ecological sanity.

However, herein this particular essay I would like to focus on what is perhaps THE key indicator which should be monitored in order to travel toward sustainable community and a neo-Earth, i.e., embodied Human Appropriated Net Primary Productivity (eHANPP). It is basic, holistically profound and comprehensive, and an indicator for determining if one's livelihood and lifestyle is socially just, humane, ecologically sane, i.e., sustainable, moral, and ethical.

As with many sustainability indicators, quantifying eHANPP somewhat precisely and accurately is generally complicated and difficult. However, eHANPP data can provide a composite focus on the impact humans and various sectors of human populations are having on the natural resource base and on life systems which are conducive to quality life for humans and associated life forms.

Units for energy used in communicating about eHANPP are joules (or one could convert joules to calories or British Thermal Units/BTUs). Moreover, remember from your biology classes or general reading, listening, and viewing, that life is matter (carbohydrates, lipids, proteins, nucleic acids, cells, populations, communities, etc.) organized with inputs of energy (or calories).

And for the most part energy to run the world (Nature's Economy and its subset, the Human Economy) is from daily solar energy and the daily solar energy of the past, i.e., savings-in-the-bank which we call fossil energy. Daily solar energy is mostly transformed into a useable form by what are for the most part the only real producers on Earth, or Eaarth, i.e., plants and other photosynthesizers.

What these unique producers (plants, autotrophic protists, chromists, and blue-green algae) capture in terms of energy in a particular area, in an ecological community of biota, in an ecosystem, or across the landscape of a nation or an ecosystem, is called primary productivity (PP). PP could be expressed as energy; however, since photosynthesizers are capturing solar energy in chemical bonds of chains of carbon (biomass), scientists generally measure carbon, biomass, or dry plant matter to estimate PP.

Embodied Human Appropriated Net Primary Productivity
eHANPP

Humans *de facto* taking more than deserved which results in serious social and ecological disparity and leaves less and less for other species, and less for maintenance of biodiversity and dynamic homeostatic symbioses ("nature") as we have known it.

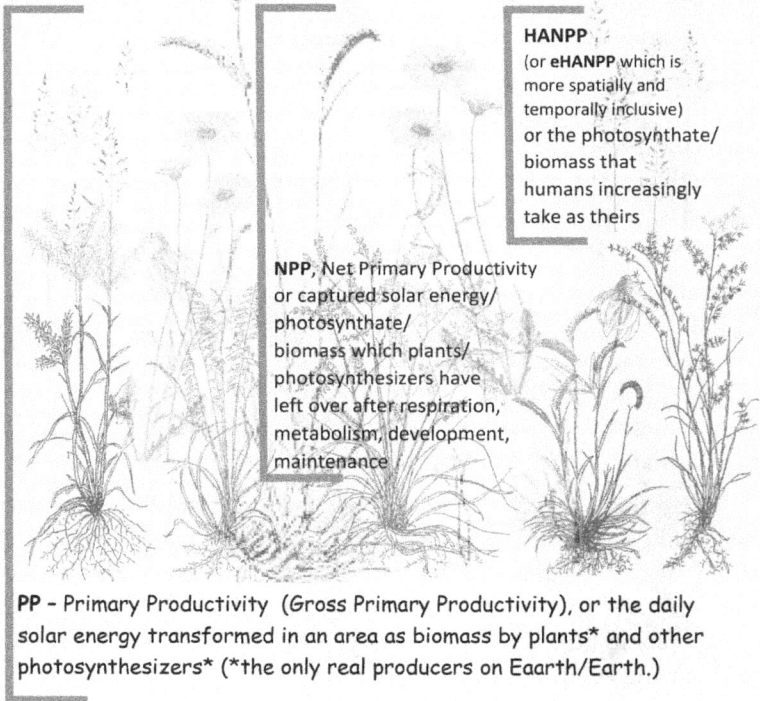

HANPP (or **eHANPP** which is more spatially and temporally inclusive) or the photosynthate/biomass that humans increasingly take as theirs

NPP, Net Primary Productivity or captured solar energy/photosynthate/biomass which plants/photosynthesizers have left over after respiration, metabolism, development, maintenance

PP – Primary Productivity (Gross Primary Productivity), or the daily solar energy transformed in an area as biomass by plants* and other photosynthesizers* (*the only real producers on Eaarth/Earth.)

The PP left over after the cellular respiration, metabolism, maintenance, and propagation needs are met for the photosynthesizers (or "plants") is termed net primary productivity or NPP. NPP can be utilized by heterotrophic organisms like bacteria, heterotrophic protists, fungi, or animals such as humans or their domesticated carnivores up the food chain or trophic levels.

On Eaarth, or in the Anthropocene, humans get the lion's share of NPP. If human actions result in harvesting large amounts of food for humans and their pets, extensive areas of pavement, plowed fields and non-native vegetation, over-grazing and -browsing, clear-cut forest-land, and excessively built structures, exorbitant amounts of NPP are appropriated for human use (HANPP), and PP and NPP can be diminished. Moreover, in order to construct the machinery and provide other inputs for paving, constructing, fencing, sawing and logging, etc., materials are appropriated from the targeted land areas as well as from other parts of the globe over periods of time (eHANPP).

What I would encourage one to do now is to take a minute and walk out across your local landscape and view the amount of paved area, extravagant homes, business structures, automobiles, plowed fields, pastures with cattle, other grazers and perhaps browsers, and/or deforested area. Now think about how the local human population is affecting NPP. Despite (and sometimes because of) the *Endangered Species Act, CITES, EPA* regulations and *Superfunds; WWF, Greenpeace,* the *Nature Conservancy,* the *EDF,* the *NRDC,* the *Tuskegee & Land Institutes,* old *LISA* and more recent "modified-*LISA*" *USDA* initiatives, the *Gates, Buffett, Kellogg, Ford, Rockefeller Foundations,* and *Master Naturalist* programs, "organic" agriculture, "renewable" energy development, innovative "sustainability" entrepreneurs, etc., etc., etc. we continue to rampantly increase our appetite across the globe for NPP, with end results of there being little of the life-enriching NPP for the relatively powerless, impoverished, landless, and disenfranchised (undocumented) refugees, or for poor humans and other species.

For full disclosure, I do wish to say that inputs of fossil energy and limiting factors such as water and key minerals (N, P, K, S, Fe, etc.) can increase PP and NPP. However, these local or regional actions generally affect global PP and NPP detrimentally and/or can change the globe such that it is not conducive to quality life for humans and biota associated with humans (i.e., resulting in loss of productive water-sheds/natural foodsheds; dead zones at the mouth of major rivers, especially in the developed, very industrialized parts of the world; global climate change; acidification of the ocean; bleaching of coral reefs; etc.).

Moreover, I need to emphasize herein that there are many sustainability indicators and ways of assessing sustainability in addition to eHANPP, which have been proposed by various psychologists, socio-political/economic researchers and other ecologists, i.e., scientists studying various aspects of PEACE/Positively Ethical Applied Community Ecology. It is just that I consider eHANPP to be the key indicator for staying on track toward sustainability for all.

Now, even though we will deal with appropriate actions toward sustainable livelihoods and sustainable community in illustrations and other narrative in this book, let's give some thought at this point to what some of the major actions to reduce eHANPP are:

1. Learn about PEACE principles, processes, and values through PEACE across curricula and campuses of all human organizational entities.

2. Learn to curb our appetite for stuff and inappropriate kinds of and excessive amounts of food, fiber, shelter, travel/transport, conventional air-conditioning, information, electronics, technology, engineering, recreation, and entertainment. Food is the largest component of your ecological footprint. Compost and raise your own food. Eat local. Primarily walk, use a bicycle, and travel by train and bus. Wean yourself from air-conditioning, a large home, and automobiles. Try to never travel by car simply to "go and exercise".

3. Cut manufacture of armaments, arms, and military spending.

4. Regulate in various ways toward curbing consumption.

5. Tithe+ and share with the truly poor in the world.

6. Reparate, give welfare and health care assistance, and "act affirmatively" when we should. And realize Land reform. Remember, if it doesn't work for the poor, it isn't sustainable.

7. Seriously consider capping income and curbing the accumulation of capital/assets. If we have power over people and resources (or NPP), we'll use it. If we don't, we won't!

And now I reiterate:

If it doesn't work for the poor, it isn't sustainable.

If we don't have an excess in power, we won't be able to use it excessively.

If we don't, we won't!

....

This little book on PEACE is not so much about the insurmountable, even though challenges exist which may seem so, and the Catch-22 Human Economy, even though this exists. It calls for a much needed "idealistic" realpolitik. It is a *grito de "¡Sí se puede!"* which deals with the Why?? What? And How??? of the process, or of our journey, primarily by utilizing illustrations, poetry and short stories.

Live lightly on the land and in
concert with Nature like we do!
Sufficiency over efficiency!!!!

We all must begin living *sabiamente*, simply,
smally, slowly, steadfastly, **SHARINGLY**,
. . .SUSTAINABLY!

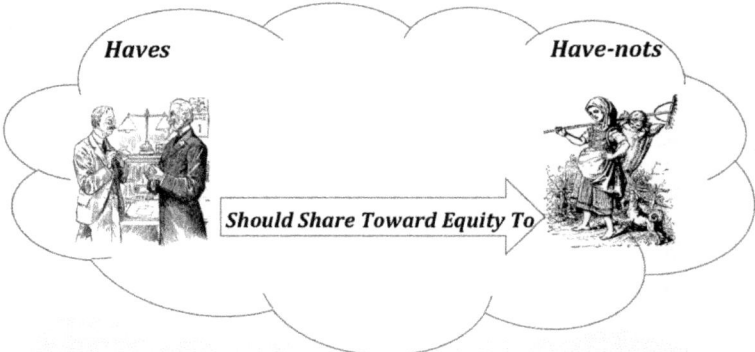

Haves

Have-nots

Should Share Toward Equity To

"To live, we must daily break the body and shed the blood of creation.
When we do this knowingly, lovingly, skillfully, reverently, it is a sacrament.
When we do it ignorantly, greedily, clumsily, destructively, it is a desecration.
In such desecration we condemn ourselves to spiritual and moral loneliness,
and others to want." Wendell Berry

. . . .

The key to following Berry's sage advice for having holistic quality community life is through
humility and a very light ecological footprint, and by being extremely cautious and tentative
in all policy and actions, **EXCEPT** in sharing with others. (These sorts of guidelines for a good
life for all, i.e., for resilient/sustainable community, have been an essential part of an ethical
and moral human ethos since *Homo* became *sapiens*.)

Solidarność?

Solidarność. Solidaridad. Solidarity. Solidarity. ... Solidarity!!!

Can we all agree that we want a better world for all--psychologically, socially, ecologically--and unite in some sort of imperfect solidarity toward that end?

A solidarity with Lech Wałęsa, Pope Francis, Dalai Lama, Jimmy Carter, and Tegla Loroupe, and with family and dear friends, in my case: Elizabeth Florence Hoffmann Martin, Louise Katherine Kneuper Martin, Kazimierz Józef Wiech, Rafael Ojeda Suárez, Darryl Lynn Birkenfeld, Marvel Jay Maddox, Luiz Otávio Campos da Silva, Miguel Angel Altieri, and many others. ... A solidarity with those who are easy, and those who are impossible??

Solidarity with the lovely people of Emanuel African Methodist Episcopal Church, the slain and the living and with the various victims of so many senseless mass killings in the U.S. and all over the world.

And yes, solidarity with the family and friends of Dylann Roof as the judge suggested in 2015 at the bond hearing in Charleston of this accused murderer. (I do agree this was a very inappropriate time and manner in which to try and persuade toward this action of appropriate solidarity!)

Solidarity with the "good" and the "evil;" with conventional capitalists and socialists; anarchists/libertarians, fascists; environmentalists, ecologists, and Greens; with advocates for PEACE (positively ethical applied community ecology) and advocates for war (convincing them that peace and PEACE is preferable to war).

Solidarity with the rich and powerful in an effort to make them less rich and powerful and with the poor, disenfranchised, and not so powerful toward enrichment and empowerment.

A loving solidarity with the crazies of the Council of Conservative Citizens, white power skinheads, NRA, Tea Party, Likud, ISIS, Al-Qaeda, Boko-Haram, (both the sane and insane within; and yes, with locos of the left as well as the right, i.e., not so right) with a goal of non-violent change toward a better world for all.

I seek a simple, insignificant solidarity with, and love and good hopes for all leaders of the world, who along with the rest of us are also simple and insignificant. Some are truly good statespersons with wonderful love for all peoples and the Nature they are nested within but also with sad, disappointed understandings of the complex and insane cruelties of Eaarth.

Solidarity for PEACE and non-violence ... and quality life for all. ... Humble, yet aggressive, solidarity.

....

Enigmas

Pablo Neruda (Translated by Robert Bly)

"You've asked me what the lobster is weaving there with
    his golden feet?
I reply, the ocean knows this.
You say, what is the ascidia waiting for in its transparent
    bell? What is it waiting for?
I tell you it is waiting for time, like you.
You ask me whom the Macrocystis alga hugs in its arms?
Study, study it, at a certain hour, in a certain sea I know.
You question me about the wicked tusk of the narwhal,
and I reply by describing
how the sea unicorn with the harpoon in it dies.
You enquire about the kingfisher's feathers,
which tremble in the pure springs of the southern tides?
Or you've found in the cards a new question touching on
the crystal architecture
of the sea anemone, and you'll deal that to me now?

You want to understand the electric nature of the ocean
    spines?
The armored stalactite that breaks as it walks?
The hook of the angler fish, the music stretched out
in the deep places like a thread in the water?

I want to tell you the ocean knows this, that life in its
    jewel boxes
is endless as the sand, impossible to count, pure,
and among the blood-colored grapes time has made the
    petal
hard and shiny, made the jellyfish full of light
and untied its knot, letting its musical threads fall
from a horn of plenty made of infinite mother-of-pearl.

I am nothing but the empty net which has gone on ahead
of human eyes, dead in those darknesses,
of fingers accustomed to the triangle, longitudes
on the timid globe of an orange.

I walked around as you do, investigating
the endless star,
and in my net, during the night, I woke up naked,
the only thing caught, a fish trapped inside the wind."

http://www.poemhunter.com/poem/enigmas/

I. TRUTH

Communicating truth??? ... "More Monnneeey!"

"He was an old-time cowboy, don't you understand
His eyes were sharp as razor blades his face was leather tan
His toes were pointed inward from a-hangin' on a horse
He was an old philosopher, of course
He was so thin I swear you could have used him for a whip
He had to drink a beer to keep his britches on his hips
I knew I had to ask him about the mysteries of life
He spit between his boots and he replied
'It's faster horses, younger women,
Older whiskey, and more money'"

Tom T. Hall*, 1975

https://www.youtube.com/watch?v=vnvMcX95G20

....

*Many of Tom T. Hall's songs are great speakers of Truth.

TRUTHS are:

1.We are OVER CARRYING CAPACITY for humans and domesticated species.

2. We are DRAWING DOWN ON THE NATURAL RESOURCE BASE.

3. There is tremendous DISPARITY ON EAARTH, and there is no way an Earth can carry 7-10 billion humans at the levels at which U.S. citizens and other Haves consume.

4. Therefore, we should all be striving to realize sustainable livelihoods and to be living "*SABIAMENTE*, SIMPLY, SMALLY, SLOWLY, STEADFASTLY, SHARINGLY, SUSTAINABLY".

5. To realize this, we need positively ethical applied community ECOLOGY ACROSS CAMPUSES AND CURRICULA of all human organizational entities (family units, schools, small and large businesses, farms and ranches, churches, government agencies/bureaus, NGO's, clubs, sports teams, etc.)

6. I do have to add that religious reliance on "invisible and visible hands" of Gods; capitalism and local, regional, and global markets; inheritance; and yes, "the government" and NGO's can be serious barriers to restoration/regeneration and conservation of resilient, sustainable ecological community. (Moreover, worship of arms and armaments, the military, soldiers, and police is a major barrier!)

7. We must advocate and fight for restoration, regeneration, and conservation of a resilient, sustainable, ecological community in every way possible. A few might listen and learn, and hopefully numbers of positively ethical applied community ecologists will grow exponentially and/or out of this process will come informed, educated, intelligent, moral and ethical, and can-do leaders of integrity and perseverance. (Although we do need to work hard for social justice, humaneness, and ecological sanity, i.e., sustainability, we need to take care not to get overwhelmed and burned out, and to do work we truly enjoy and to bite off what we can chew.)

....

We've been screwing up dynamic homeostatic symbioses/"nature," realizing genocide and enslavement and destruction of sustainable social fabrics/systems, losing rich irreplaceable top soil, clearing our fire-dependent southern long-leaf pine-dominant forests and short-grass prairies and other sustainable ecological communities, polluting and depleting waters and the healthy biota in them, depleting biodiversity, increasing embodied human appropriated net primary productivity, for a long time now in this old world. In many cases we have temporarily band-aided these problems here on this Eaarth with technological fixes and associated fossil energy inputs but put off real solutions to another day for other generations "to deal with." We really have not truly begun to sustainably confront agrilogistics, industrialization, and this current world of electronic/electromagnetic data and information rather than one of knowledge, wisdom, and prudence.

....

I do wish to say "I'm sorry" to all good folk who fight for a change toward holistic recycling, no-till agriculture, organic farming, renewable energy, Green New Deals, saving small rural communities, etc. I do generally support these efforts in hopes that they might serve as a quick and relatively immediate transition to real solutions.

However, (Again, I apologize to you good folk with much "positively ethical applied community ecology" in your heads, hearts, and souls.) these little superficial efforts must result in a world of sustainable livelihoods in which we ALL live "*sabiamente*/wisely, simply, smally, slowly, steadfastly, sharingly, sustainably." We Haves generally need to reduce our energy transformation from about 150,000-300,000+ kilocalories/capita daily to ca. 2/3rds of that … and share power with the have-nots (and other species) to get energy transformation/capita world-wide somewhat closer to equity.

Therefore, we need immediate radical, revolutionary change to our socio-political/economic systems (world-wide) if we are going to make it much longer as an existing species.

What do we need? A future EARTH of quality life for all, including other species, for as long as possible, i.e., transformation of the current Anthropocene Eaarth to somewhat as it was 50,000 years ago (Pre-Information/ Electronic/Chemical, Industrial, Agricultural ages)

good's creation

SOLIDARITY

Positively
Ethically
Applied
Community
Ecology

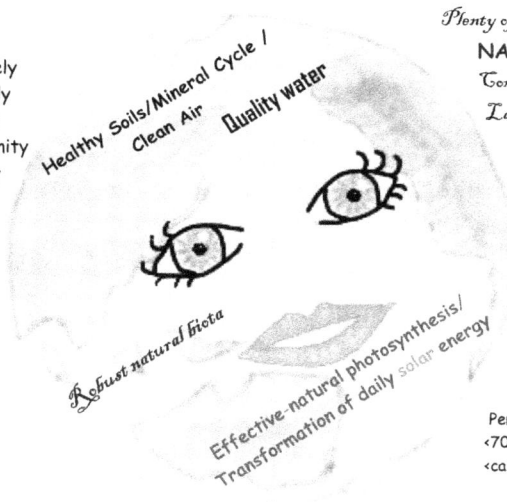

Healthy Soils/Mineral Cycle /
Clean Air

Quality water

Plenty of:
NATURE (symbioses)
Commons,
Land

**Ethical
Energy**
Transformation

*Universal
Human
Humility*

Robust natural biota

Effective natural photosynthesis/
Transformation of daily solar energy

Per capita equity of:
<70,000 kilocalories/day;
<ca. $50,000/yr.
(2020 U.S. $)

Sustainable Livelihoods
(Living *Sabiamente*, Simply, Smally, Slowly,
Steadfastly, SHARINGLY, SUSTAINABLY)

Beauty, Energetics, and Limits on Quality and Quantity of Beauty

"Beauty is truth, truth beauty,'—that is all
Ye know on earth, and all ye need to know." Keats

But what kind of beauty? And how much of it? Moreover, what is the energy/emergy involved, and what effect does it have on symbioses/"nature"/creation?

The clear-cutting of forest can result in beautiful open meadows. The contours of agricultural beds prepared for planting of domesticated seeds are beautiful. Ordered white-fenced, red-barned, neatly-painted clap-boarded and shuttered rural communities of flower and vegetable garden landscapes can be very, very beautiful. Ecological communities set back in productivity by disturbances of bulldozing, plowing, mowing, burning, or other perturbations are often transformed into beautiful displays of native wildflowers. All 7 billion humans, their zygotes and embryos, their ova and spermatozoa are beautiful. Domesticated breeds of, for example, horses, cattle, cats, dogs (and even the show animals) possess beauty. Frank Lloyd Wright-type architectural feats in the built environment are structures of beauty. There is beauty in the skylines of major metropolises. Beauty can be schematic illustrations of all the crazy lines of communication across the globe, the flyways of migratory birds, or the migratory patterns of *Homo erectus, Homo neanderthalensis,* and *Homo sapiens.* Fires can be very beautiful. The immense explosions of nuclear bombs are awesomely beautiful. The blasted sands out from Los Alamos exhibit beauty.

Beauty is a function of mindset. A yard of "weeds" struggling through secondary succession can be beautiful and should perhaps be considered more beautiful than mowed and trimmed lawns of non-native bermudagrass, zoysia, or centipede grasses. However, advertising, traditions, actions that are for realization of safety and security, and human evolution have generally led to most humans preferring the manicured landscapes over "natural" secondary, successional, ecological communities, or even "climax" ecological communities.

Beauty can be clearly destructive and even evil, as is definitely the case for some of the examples in the second paragraph herein. Moreover, beauty can result from or be utilized by propaganda machines with evil intentions and/or with very negative results. Finally, too much of anything on an Earth that has limits in its natural resource base and amounts of net primary productivity is destructive to symbioses ("nature").

With an Eaarth population of seven billion humans, which will reach perhaps ten billion, consuming in some instances more than 300,00 kilocalories/capita/day (among the Haves), and appropriating much more than a lion's share of net primary productivity, beautification via rampant artificialization has gotten out of hand! We need to slow down as humans and begin to truly appreciate the beauty of symbioses.

You can't beat "natural" fractals for true beauty. The beauty of homeostatic symbioses is Truth. And truth is natural beauty. That's the truth! Therefore, that is all one needs to know to begin the process of transforming Eaarth to Earth.

....

The Beautiful Dream of the Cosmos and Life in the Cosmos Involves Fractal Systems from

the Big Bang   to   Eternity

(which can be analyzed & modeled using statistics and algorithms).

....

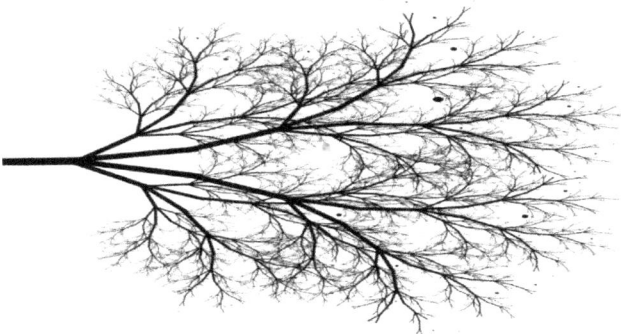

The deep and real "natural" beauty of dynamic homeostatic symbioses also has everything to do with energetics. If the beauty is not capable of being sustained, restored or regenerated through daily "natural" solar-generated photosynthate or with relatively little embodied human appropriated net primary productivity, or if the ecological footprint involved in producing the beauty is too large, or if the technology employed is inappropriate, then "the beauty" isn't innately, profoundly, and sustainably beautiful.

A discussion of energetics is complex and confounding. In terms of sustainable payoffs from energy inputs, throughputs, and outputs, one must consider:

- what diffuse, concentrated, and embodied energies are involved,

- what are the negative externalities from the processes of energy transformation,

- what increased inputs in a subsystem might do to homeostatic symbioses which coevolved for millions of years, and

- what particular human uses of the energy result in overall good.

The appetite of one collaborator on this book, paul bain martin, for at least a minimal understanding of energetics was first whetted in 1971 by the gracious, knowledgeable, persistent, and wise ecologist, David Pimentel of Cornell University, who had kindly sent paul numerous reprints of publications on population dynamics and biological control of the lepidopteran herbivores on which his doctoral research focused. Later, triggered by the energy crises of the 1970's, Dr. Pimentel became involved as a pioneer in studies of energetics of agricultural systems and was always ready to patiently explain some aspect of this important agroecological topic when paul called or wrote him with a related conundrum. Then in that same time period in the community ecology classes of turtle-migration expert Archie Carr, paul learned of and began to truly appreciate the Silver Springs research and systems modeling investigations of Dr. Howard T. Odum involving energy/emergy transformation and flux.

In one of the Odum short-films which collaborator paul presented to his principles of biology and environmental biology classes at St. Philip's College, the interviewer put H.T. on the spot about using a car to travel to the location where the interview and filming were done versus more appropriately walking or bicycling. (*https://www.youtube.com/watch?-v=6I7zcYyomyA*). In addition to H.T., paul has read that object-oriented philosopher/"ecologist," Timothy Morton, and environmentalist, Bill McKibben, have been criticized for using heavily fossil energy-dependent and "unsustainable" modes of transportation such as flying to research sustainability and to disseminate information and knowledge related to this topic and advocate for sustainable livelihoods. In addition, paul was criticized by students for using a motor vehicle to a drive 33 miles to St. Philip's College in order to make the case in classes about how inappropriate automobiles are in transport systems.

We are all sinners; no one is perfect. But above and beyond this is the fact that Truth and processes toward Truth are messy, and at times we must create chaos in time-space subsets of symbioses in order to get to a more symbiotic order holistically and to realize sustainable livelihoods for all.

Dr. H.T. Odum's films and other works have been things of deep beauty in terms of contributing to our knowledge. Nevertheless, in his process of developing that ordered knowledge and in disseminating it toward knowledge, wisdom, and prudence for all, energy has been transformed, and some entropy and chaos have resulted per the Second Law of Thermodynamics. And despite the entropy created as a result of their beautiful creation of ordered knowledge, the works like those of Helmut Haberl and colleagues at the Institute of Social Ecology in Vienna, Austria, toward consilience and a better undertaking of applications in healthy social ecology will help us toward realizing more Truth and good Beauty even while it is a very messy process and will result in some messy outcomes.

Why? Eaarth/the Anthropocene. The mess we have made and are in.

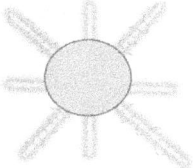

Yep!. To a large extent it was the hubris of agriculture, the industrial revolution, and the synthetic chemical/electronic/information age which caused this.

Artificial: powerful humans-centric (We act as if: "We are the world!")

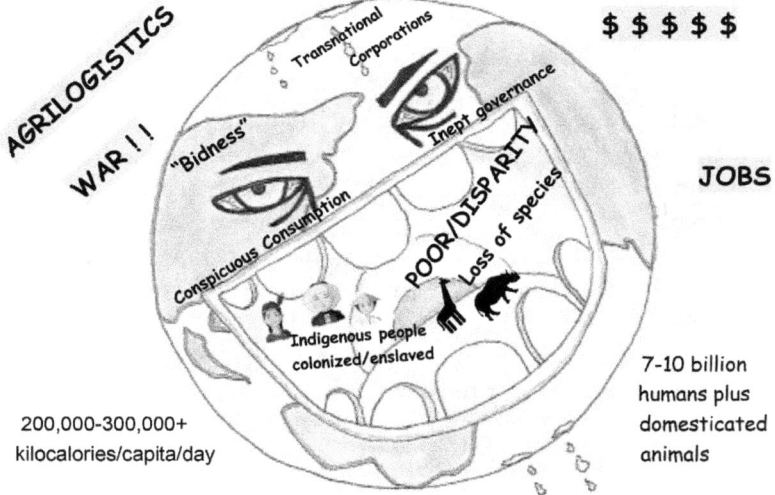

AGRILOGISTICS

WAR!!

"Bidness"

Transnational Corporations

Inept governance

Conspicuous Consumption

POOR/DISPARITY

Loss of species

Indigenous people colonized/enslaved

$ $ $ $ $

JOBS

200,000-300,000+ kilocalories/capita/day

7-10 billion humans plus domesticated animals

[Nevertheless this very obvious artificial system (A) is but a subset of the natural system (n). If we are not aware of this, "weeds", "pests," "varmints," "diseases," "rot," "destructive weather," and geological disruptions remind us.]

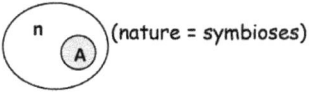

n (A) (nature = symbioses)

Armistice Day … For Ever and Ever!

I am very much a mess,
And I do know it.
And I hardly never, ever not show it.

Perhaps I should wish I were a Bible Thumper, a Christian
with a big "C,"
A believer kneeling often, chanting "Hail Marie."
A soldier, a patriot,
Comfortable standing with hand over heart
For the National Anthem, for Flag,
For You My God …
Thou Art.

But I don't
And I won't!

However, this 1960's (mess of a) child
Who thinks things are so wrong,
Will be all right,
ECOLOGICALLY
When we ALL get along!

Wait! … Let me do more with this song
Lest you feel I do you wrong.

There were many of my family and friends
In the Pacific, North Africa, and Europe …
In the wars "Over There."

A true elder, Ossie Davis' young brother,
Near and dear to our hearts
Returned from the Korean War
Shell-shocked and with wound.
Doc Davis still needs VA treatments,
Which will not end soon.

Childhood friend Robert Cruz left Vietnam
With terrible PTSD scars.
LIFE IS NOT FAIR!
(I sort of beat the system
And, as a gringo Naval Aviator, never went.
I DORed.[1])

Dad, Marine Master Sergeant Alton Martin,
With eyes and songs that were truly blue,
Was at the horrible Battle of Peleliu.

And Uncle Bain--
Blasted into bits--
In Germany remains.
This left the family and many friends
Of this quiet Clark Gable,
Of this horseman and athlete
So extremely able,
Of this beloved son of rural south Texas ...

It left his loved ones in terrible fits.

"Dulce et decorum est pro patria mori."
So much bullshit!!!
From the European Front near that old Rhineland
Uncle Oscar Bain Martin, my namesake, said it true,

---

1. This wannabe pacifist was "dropped on request" by a benevolent commanding officer and honorable discharged.

For the world, for the Red, for the White, for the Blue…
"If PEACE is planned as well as War is,
We should never have War to go through again!
We'd have PEACE forever!"
"Dulce et decorum est pro patria mori."
So very much bullshit!!!

Wait!… I'm sorry,
But there must be more
Before we sort of begin
To close the door on War.
It is known that the process is tough and lengthy

Needing
Sophisticated peaceful plans,
Peaceful strategies,
Peaceful tactics,
Peaceful measures,
Peaceful replanning,
And then new workable peaceful measures.
It's complicated,
Especially with inept leaders
Inept systems of ecological education
And inept ignorant masses.

We should fight the Might
Of status quo
And the powerful evil within.
Fight the ambush killings in our schools
De facto facilitated by the NRA's
And Trump and Trumpists
With their tweeting ways,
Massacres of the good in our Social Fabric,
War on dynamic homeostatic symbioses,
"Nature," the environment.
We must win against misogyny, the overly ignorant;

Racism, xenophobia, tribalism;
Neoliberal capitalism, consumption, greed;
Warmongering, love of guns, adoration of soldiers and
military;
Hallowed battled grounds and hallowed military machines;
...
Against lust for Lands;
Against power-drunken, pompous, narcissistic inhumanity;
Against despots, demagogues, dastardly bad tyrants;
Against Neo-Hitlers, Mussolinis, Francos,
Stalins, and Perons.

Such terrible asses.

Wait! ... Maybe all is okay?
God, the invisible hand of neo-liberal capitalism;
Trump, Duterte, Putin, Jong-un, al-Assad, Bolsonaro,
        Erdogan
Will lead the way.

They'll have us go for the Gold,
Fight for the tribe,
Pro patria,
And we, the masses, can complacently and apathetically
Close our eyes, plug our ears, resist the temptation to feel,
Continue to be brain-dead,
Marching onward *pro patria*
With our guns and our bombs,
Our gas and germs,
Our words—
Nasty infectious evil propaganda—
From indoctrinated, pent-up, bombastic,
Hateful hate of haters
Behind our Divine, Holy, Blessed, Prayed-over
And all-knowing leaders.
We and Thee

Can doze away with Prayer
And Be Blessed.

Deep down few truly believe
This warped bullshit.
And though some thoughts do go to despair
Each and every day—
We know a profound, comprehensive, and holistic Golden
Rule;
The humble, gentle, peaceful, and ecologically educated
Maintenance of our genes;
Hope and charity; and
PEACE;
Armistice for ever and ever
These are The Way.

Is Devine Divine Anymore?

It Was Far from Being Perfect. But, What Happened to My Wonderful Home Just East of Devine, Texas??

The Warm & Fuzzy Feel* and Sustainability of Nature and the Land... Has Yielded to the Rampant & Cold** Artificialization of Asphalt, Concrete, and the Built-Environment (and Yes, even Walmart!)

(*But also, with plenty of heat, drought, and stickers, goatheads & thorns!! **"Cold" even if it is giving us global warming!)

paul bain martin

In my formative years just outside the city limits of Devine in the 1950's & 60's, my Alton and Louise Martin family owned less than five acres on which at any point in that general time period we were raising hogs, a milk cow and calf, and chickens and maybe a few guineas. We were tending to a large garden area as well. However, we never felt restricted to these five acres.

From an age of about five or six years my five siblings and I had the freedom to roam, hunt, run, and swim in my uncle's irrigation reservoir and on over one hundred acres of native pasture (under secondary succession) immediately around us, tracts of which belonged to our Uncle Peggy Martin, Mr. Pete Gutiérrez, Mr. Fritz Schroeter, and Mr. Fred Bowman. From our common ages of six to eight years, in those pastures Esteban and Alejandro Peña taught me uses of wild plants and Spanish names for wild animals. Their dad gave us gallons of honey, which the busy social critters from the Old World had made from the abundant horsemint growing on Mr. Gutiérrez's and Mr. Bowman's land.

We absolutely loved that land and holistically learned and dreamed on it. Moreover, it was an environment in which to positively deal with

86

our challenges of family, school, and the local human population; melancholy; or the craziness of a very crowded non-airconditioned household of two bedrooms, a very small bathroom, and eight people.

A kid of 2020 being raised in that same locale would not be able to realize the freedom and good life we were provided. What happened?

Neo-liberal capitalism, fast-paced commerce and development, and Eagle Ford Shale fossil energy activity has changed the area from savanna pasture and small farming/relatively low-input agriculture to Interstate 35; big air-conditioned pick-up trucks, semis, and other automobiles laden with electronic devices; businesses; air-conditioned housing with virtual realities from electronic devices; fossil-energy-produced foodstuffs, which are mostly brought-in-from-afar, and lots of it!; and a considerable amount of built-environment of asphalt, concrete, and imported materials that comprise artificial structures in Devine.

However, thorough explanations for "truths"/realities are not easy, are always complex, and must always involve a sort of chicken and egg cycle. It was largely changing mindsets and behaviors in conjunction with dramatic landscape deterioration towards the artificial and more built-environment, which happened to my home in Devine, mostly in the first seventy years of the twentieth century and especially immediately after World War II

Both dads and moms have been working to have more, More, More!, More!!, MORE!!!, and they often have left their kids to the care of others in settings not as conducive to real freedom for these youth.

Play, recreation, and learning have increasingly become a plastic indoor world of Legos and other toys, along with the virtual realities of televisions, computers, and small handheld electronic information gadgets, which focus on the artificial and virtual rather than the natural and real.

Many parents have become "helicopters."

We are all very artificial and unnatural, and we are distant from each other in our local, ecological communities. We have lost a sense of place, ecological community, and natural spirituality. Families, neighbors, and local human demes or populations have lost intra- and inter-connectiveness and commonalities and do not have empathy and compassion for and do not know, respect, trust, and genuinely love each other.

Kids are not as able to be free, become little ecologists, and develop critical thinking skills anymore! … However, who really cares about our progeny and Nature when we have air-conditioning, Triple-C steaks, lots of wonderful plastic and new technology, and Walmart?!

....

As I have indicated before, there are more stories like this in my 73+ year-old head which come from a time on Eaarth of much artificialization. There was an anticipated joy of returning late one night with a student and camping in Oasis State Park, Portales, New Mexico in the middle of a high plains/shortgrass prairie ecoregion. Thirty years before my wife and I had visited this relatively pristine area. The joy experienced by my student and I turned to disappointment the next morning as we woke to the smell and landscape of the large Holstein Friesian-dairies, whose owners had fled regulation in California and had settled around this lovely site. Another such disappointment involves my wife's grandfather's Hoffmann Ranch, west of San Antonio where several of the Texas ecoregions come together and where my wife enjoyed her early years of growth. It is now covered with houses of the Alamo Ranch subdivision.

We? Or Me? ... Or WE!

(of Dynamic Homeostatic Symbioses)

"We are much better [off] than we were 250 years ago."
Spoken by a friend I dearly love (in spite of his Eaarth[1]-
denial).

....

"We ... (are much better [off]")...
Is ME!

"'I' am much better [off]."...
Not WE.

Not those of potentially-wonderful, life-teeming organic
    topsoil
Covered with asphalt and concrete
And massive edifices (of capitalism),
Smashed to chaos by billions of automobiles.
(And smashed/overwhelmed
By human beings/
By domesticated species.)
Destroyed with exotic, non-native species,
Steel, plastic, synthetic hydrocarbons;
Disrupted and poisoned by industrial agrilogistics[2];
Virtually ignored by trans-local corporations,
IT, "the" internet, neo-liberal capitalistic algorithms.

---

1. Bill McKibben, Timothy Morton/(Jared Diamond), Tim O'Brien/Darrell Scott,
Molly Ivins, Wendell Berry, Muhammed Ali, respectively.
2. From Rice University philosopher, Timothy Morton.

Not oceans choking on plastic,
Poisoned (along with lipids of humans and other species)
With DDT, DDE, DDD, and
Other toxic and hormonic-mimic attackers
Of nature/
Of dynamic homeostatic symbioses.

Not the energetically-sound southern long-leaf pine
Climax-ecological communities,
Clear-cut and developed.

Not the dynamic homeostatic Oglala prairies,
Plowed out,
Treated with
Massive amounts of atrazine, 2,4-D,
Trifluralin, glysophate, chlorpyrifos,
And other biocides;
Mined for fossil carbon molecules
Laden with dangerous amounts
Of captured photosynthetic energy
And potential pollutants.
Mined for rapid-transformation of energy
To run dying high-input/throughput systems of human
        consumerism.
Systems of morally-corrupt decision-making,
And void of true spirituality.
Systems ignoring Truths;
Releasing climate-changing CO_2 & CH_4
At chaotic rates ...
"To keep the dirty lights on"[1]
And feed our enormous guts
And allow for our various stupid toys
And games of war ...
In every sense of the word ...
At many levels and
Amounts of intensity, social injustice, inhumaneness,

ecological insanity;
"Developed" to keep non-sustainable
Urban, and rural communities
On opiate-needles of destructive
Capitalistic "bidness[1]."

Not steadfastly proud hunter-gatherers
Of indigenous reservations
And other locales
Across Eaarth.

Not many individuals, demes, populations
Of homeless and others marginalized
In various circumstances
Of socio-economic, political injustice,
In favelas, slums, prisons/
ICE detention and other impoverished areas,
Often involving hate and discrimination and bigotry,
Against skin pigmentation, ethnicity, sexual orientation,
        etc., etc.,
In various sectors of Eaarth.

Not the passenger pigeon, western black rhinoceros,
Tasmanian wolf, great auk,
And thousands and thousands
Of other species
Of all Domains and Kingdoms
Of the Anthropocene.

Not the poor creatures of true and real living-systems
Who have been chaotically-destroyed
With all this artificiality.

    ....
It's a zero sum game.

We're in this together.
All must live according to Wendell[1]:

> "To live, we must daily break the body and shed the blood of Creation.
> When we do this knowingly, lovingly, skillfully, reverently, it is a sacrament.
> When we do it ignorantly, greedily, clumsily, destructively, it is a desecration.
> In such desecration we condemn ourselves to spiritual and moral loneliness, and others to want."

Ali[1] once spoke
A powerful piece of poetry:
> "Me? ... We!"
Not war on other biota of our and other species.
Not war on ecological communities.
Not war on dynamic homeostatic symbioses.

Think humility,
Justice, empathy.
Live *sabiamente*, simply, smally, slowly,
Steadfastly, sharingly, sustainably.
Think/Rock[2] in solidarity and concert
With all of nature.

Yes!!! Yes!!!
Yes.
I too use the opiates of corporate industrial Eaarth,
And abuse.
Still, in all my frailties and limitations,
I seek to never cease to criticize
And fight (hopefully somewhat intelligently and
    systematically)
For dynamic homeostatic symbioses.

And I AM SORRY.
I do not appreciate self-inflicted ignorance.
Apathy, smug complacency,
Ignorant, narcissistic arrogance,
Blessedness.

If We of Eaarth are not better off,
I am not better off.

WE are not better off.

Why? What? How? The Truth Process toward Quality Life. Facilitating toward a New Earth.

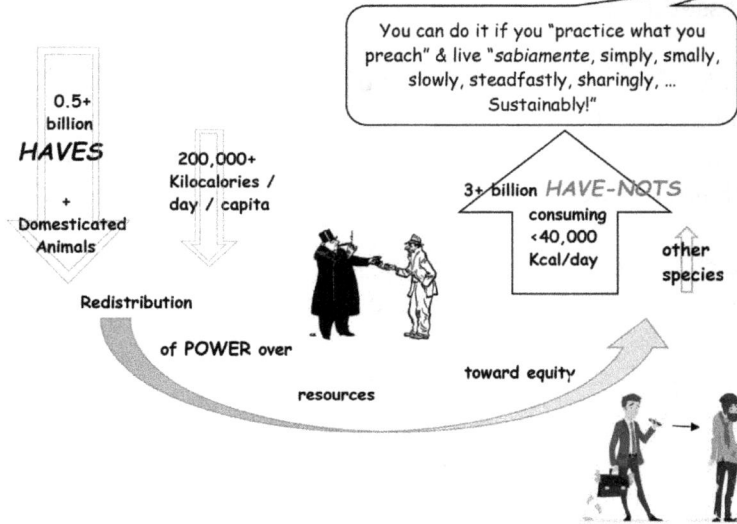

You can do it if you "practice what you preach" & live "*sabiamente*, simply, smally, slowly, steadfastly, sharingly, … Sustainably!"

0.5+ billion

HAVES

+

Domesticated Animals

200,000+ Kilocalories / day / capita

3+ billion *HAVE-NOTS* consuming <40,000 Kcal/day

other species

Redistribution

of POWER over

resources

toward equity

Ethic of Reciprocity/Golden Rule (toward all … species)

Abide by the Precautionary Principle & 2nd law of Thermodynamics

Health care/nutrition, education, water, Land, housing … as a right & not owned/for profit

Open borders, but with care for Have-nots & other species

Ecological literacy (ecology across curricula/campuses of all human organizational entities)

½ of Anthropocene Eaarth to Nature (returning it to Earth)

Appropriate applied agroecology

Real communication, respect, empathy, compassion, humility

Local food/fiber/shelter (& including low-input entertainment, recreation/tourism)

Walk, bicycle, use mass transport

Live: *Sabiamente!* Simply! Smally!! Slowly! Steadfastly, and Sharingly!!!
SUSTAINABLY!!!!!

Thoughts on Juneteenth, 2019

Whether you became a U.S. citizen yesterday or were born a citizen 120 years ago, slavery is one of your original sins. And we must seriously consider and use where appropriate processes and tools of

- regulated capitalism,

- democratic socialism,

- affirmative action,

- reparations[1], and

- local and federal governmental policy/laws/assistance

to move us toward ecological (PEACE) learning and appropriate infrastructure enabling sustainable livelihoods for those put in a cultural, socio-political, economic rut because of slavery (which in very many ways still exists today).

---

1. Coates, Ta-Nehisi 2014 The case for reparations, The Atlantic, June Issue

Why? Challenges/Barriers for Changing Eaarth to Earth. Anti-science and conspicuous consumption.

The big challenge is from the Religion of capitalism/materialism/consumerism to the Scientific reality of a carrying capacity of 2-4 billion Haves, global (climate) modification of Nature by humans, and biophilia.

Fast Pace & Loss Of Sense Of Place / Community

NOT LIVING IN THE MOMENT

Cement / Concrete

Large Overly Organized Homes

Conventional/Current Capitalism/Neoliberalism

Glamor

Efficiency vs. Sufficiency

CLOTHES

LIVING IN BUBBLES

Blind Faith

Little Critical Thinking

Resistance to Change

Haber-Bosch

High Input Entertainment / Recreation / Tourism

Big Pharma

Population Density/Numbers

Large Manicured Lawns

Plastic

BIG SCHOOLS

STUFF

Fossil Energy Addiction

CARS

Paucity Of Statespersons/<u>Real</u> Citizens

Poor Sustainability Indicators

Conventional Agriculture (Including Big Organic)

Too Much Emphasis On ***Recycling vs. Reduction***

Ecological Ignorance

Con-Men/Advertising/Bad Propaganda

VIRTUAL REALITY

Football, Pro Sports

World Wide Internet/Androids/etc.

"Faster Horses, older Whiskey, more Money"

SURVIVAL

Entrepreneurs

Pseudointellectuals

Clutter

Inappropriate Technology / Engineering

A Sick World of Plastic

It is essential that we ban disposable plastic bags (and many other uses of plastic) and begin rectifying at least some aspects of our downright sinful ways. I do wish to make it clear from the outset, that despite these anti-government, libertarian times, places, and people, and despite the prevalence of an anti-science sentiment, this particular human has no problems with using science, learning, carrots and sticks, and overt and covert behavior-modification in the fight to protect our Commons. Protection from plastics of the Commons, the Land, and Nature and the natural resource base of fresh and marine waters, soil and air (e.g., from pollutants from manufacture and other aspects of the life cycle of plastics), and healthy biodiversity including daily solar-energy-transformers/or photosynthesizers is needed for regeneration and conservation of resilient, sustainable ecological community. We have a duty as citizens to educate, govern, and live in sustainable livelihoods toward a clean land and natural environment of excellent quality for all humans and other organisms, in the present and the future.

As a Catholic, let me confess to some of our sins which are not necessarily plastic, but, nevertheless, associated with plastics, as is almost everything in today's world. "Bless me Father for we Haves have sinned considerably through the following behaviors:

- Eating and drinking too much, and exotically,

- Drying clothes other than on a line,

- Using cars,

- Removing leaves and debris with a leaf-blower,

- Building large air-conditioned homes with exotic landscapes,

- Flying and taking fossil-fuel guzzling ocean and river cruises, and

- Going to war to hang on to our power or grab more."

I do fully realize we are very flawed genetically and epigenetically as a species, and that this will not change significantly in the future. However, this U.S.A. and international system of governance, laws and regulations, and education could be structured such that it steers us toward an ethos, values, and behaviors, which would result in a better world, a neo-Earth, for all of us and for our progeny. We could and do blame the Donald Trumps of the world, and local, state, national, or international politicians, leaders, businessmen, and a variety of manipulators for our sins. However, ultimately, it is we ourselves who are to blame, and we ourselves must begin to individually and collectively initiate and realize real change at local levels.

But let me get back on track somewhat and focus on disposable plastic bags. It makes no sense to depend on these serious polluters of our oceans and marine biota, significant transformers of quality energy, and disruptors of natural systems. Most of us can lighten our footprints and help to regenerate and conserve a healthy Earth by swearing off disposable plastic bags.

For transporting purchases of food and other necessities, use washable cotton bags. Purchase fresh or dried foodstuffs (grown locally if possible, including around and in your own home) in bulk and prepare them in a relatively efficient manner, avoiding use of electrical heat. Store much of your food in washable and durable glass.

And, you/we can do it! During my years with my Louise and Alton Martin family in my home area in Devine, Texas in the 1950's and 60's, we did use some paper bags for groceries and other purchased products. However, to a large extent we adhered to the behaviors and practices recommended in the previous few paragraphs primarily because disposable plastic bags (and indoor automatic clothes dryers) were not readily available, and because frugality, reuse, and local foods were the norm. For many of the same reasons, while living in Brazil in the early 1980's, our young Betsy and paul martin family of five for the most part did without disposable plastic or paper bags. We carried permanent bags when we were out shopping.

Nowadays, because of too much of the types of the sad and deplorable mindsets and behaviors Aggie Clayton Williams called for in his serious stumble during the 1990 Texas gubernatorial race, whether I am in Brazil, Central America, or in the U.S., the blight of disposable plastic bags can be seen everywhere. Moreover, the disposal bags are only part of the mass of plastic coming out of our stores each hour. In observing Seguinites leaving our HEB grocery store, in addition to the unbelievable amount of disposable bags being used, sometimes with but one product per bag, there is the wide variety of products contained on and off the shelf by a comparable amount of plastic by weight to the amount of actual product, which also may be mostly of plastic. (In specifically focusing on food products in our grocery stores, it is no wonder that our nutrients cost so much more in terms of energy inputs [energetically] today when compared to pre-WW II, and that it takes well over ten calories of input to produce, process, and transport each calorie of food we eat today in 2018.

Lott, Melissa 2011 Ten calories in, one calorie out. The energy we spend on food. Sci. Amer. Newsletter August 11 *https://www.cias.wisc.edu/wp-content/uploads/2008/07/energyuse.pdf*)

The current status of our plastic world, Eaarth, is absolutely sickening! Since plastic manufacturing began to take off in the 1950s, over 8 billion metric tons of plastics have been produced and are still around, and 100 billion plastic bags are used every year by U.S. citizens. Estimates are that 91% of manufactured plastics are not reused or recycled. And at the current rate of our trashing and polluting the oceans, by 2050 there will be more plastic biomass in marine environments than that of fish.

Gooljar, Jason 2018 Fact sheet: end plastic pollution. Earth Day Network, Green Cities. Mar. 7

Poem of Hope

(PRAYER FOR LIFE-SUSTAINING SOFT POWER ON EARTH)
Response to an assignment by
Generations Indigenous Ways leader, Waylon Gaddie
Pine Ridge Reservation, SD, Summer 2019

paul bain martin

Waunspe
Wacekiya[1]

Let us learn on our own.
And let us prod ourselves
To actively do so.
True learning has to be "hands on,"
Often painful.
We've got to get dirty.
But it's worth it
For all of life ...
For dynamic homeostatic symbioses.

Camarón que se duerme
se lo lleva la corriente.
The shrimp that falls asleep
is carried by the current.

(There were many Wise words
Spoken by the peoples
Of Generations Indigenous Ways
In Yellow Bear Canyon

---

1. *Waunspe wacekiya* means "pray for wisdom." (Helene Gaddie, Generations Indigenous Ways, Pine Ridge Reservation).

On a fatigued but powerful
Thursday, June 27, 2019.
We pray a prayer of Hope
For real action
That we heed,
We follow through
On these wise words.)

"Soft power is necessary."
Science educators/
Key organizers of Generations Indigenous Ways
(Pine Ridge Reservation),
Helene & Waylon Gaddie would say.
"POWER sustains life-forms and life systems.
Life is matter ordered through inputs of quality energy."
The big hurdle, though,
For sustainable life systems
Is the second law of thermodynamics.
Order in one system creates chaos in another,
And there is always a tendency towards chaos.

Therefore,
Power does sustain life,
But life is always connected
And too much power in any life-form or system
Causes chaos.
And chaos is the antithesis of life
Or the opposite of life,
Especially quality life ... for all!

And this destructive chaos is rampant here,
Especially on Turtle Island[2]
Yes, Turtle Island has become the occupied center
Of life-system-destruction,

2. Indigenous name for North America.

Relatively non-regulated growth
And a fetish-like focus
On non-sustainable measures from Wall Street &
 GNP-economists,
Via neoliberal capitalism,
Especially during this (hopefully) brief Trumpian era.
Yet, it was Western Europe
Which kicked capitalism in gear
And ripped out the heart and soul
Of the social fabric
Of Turtle Island, South America, India, Egypt, and much of
 the Middle East and Africa ... .

WE ARE ALL ARTIFICIAL, GLOBALIZED, AND PART OF THIS
GLOBAL CAPITALISTIC SYSTEM WHETHER WE LIKE IT OR NOT.

¡Y escúchame!
First listen to this example
Of one little aspect of European expansionism
And industrialization
And artificialization!!!
(I'll return shortly to this "listening" matter
Later In this poem/ prayer.
But for now
Listen to this one little point!!!)

Nitrogen is in much of the biomass,
In the protein, plant & animal,
In the ATP,
In the DNA
Of/embodied in domesticated animals.
It is in the Mississippi watershed,
In the dead zones of this Anthropocene
On this current Eaarth
(Eaarth spelled with a double "a"
To mean "Earth in the Anthropocene,"

"Earth of the Artificial").
Artificial nitrogen molecules,
Fossil energy tainted,
Fossil material tainted
Ammonia, urea, nitrates are injected
Through agrilogistics
Into artificial & natural living systems,
Initially by the ubiquitous Haber-Bosch industrial process
Using natural gas, coal, and oil,
A process which was developed in Germany
In the early 1900's.

(By the way,
Science is a human process.
And though it often yearns to be objective, pure, moral, and
        ethical,
All things are relative,
And relative to being totally objective,
Science continues to be subjective,
And like the life it seeks to better,
It is somewhat messy.)

AND THE BEAUTIFUL LEGUME-COVERED PRAIRIES?

Back to nitrogen and artificialization ...
On many of what were native prairies
And ecosystems of South Dakota,
Grazed by bison and cattle,
Ecological imperialism/globalization provided sweet clover[3],
And other exotic, invasive legumes
Which fix considerable nitrogen for all of the "natural"
organic molecules
(And other systems mentioned previously).
Globalization means that we are indeed all globalized,
industrialized, artificialized.

---

3. *https://en.m.wikipedia.org/wiki/Melilotus_officinalis https://en.wikipedia.org/wiki/Ecological_Imperialism_(book)*

AND NOW WHAT? TREAD LIGHTLY, BUT STEADFASTLY,
WITH SOFT POWER.
AND, FIGHT HARD ... ADVOCATE RELENTLESSLY FOR
ALL OTHERS TO JOIN US IN REALIZING A SMALL
ECOLOGICAL FOOTPRINT

So, what is the solution
To this conundrum,
This riddle
That life needs power,
But Power creates chaos???
SOFT power! SHARED power!
Don't capture, do not enslave, don't colonize, do not
occupy.
Don't rip the heart and soul out of culture and Nature.
Eliminate genocide and artificial extinction.
And do this through "Positively Ethical Applied Community
Ecology",
Through PEACE,
Through utilization of soft power.
Attempt to tear down all walls;
Destroy armaments, arms;
Abolish militaries.
Use resources for living,
Sabiamente/knowledgeably/wisely/prudently,
Simply,
Smally,
Slowly, ...
And, yes, steadfastly!! ...
But sharingly, (caringly), ...
Sustainably.

MORE THOUGHTS ON "THE WAY"

How do we get there from here?
Here on this chaotic Eaarth
The Eaarth of the overly artificial,
Of the Anthropocene?
(It does seem virtually impossible
To transition from a War-on-Earth mindset.
Nevertheless,
Morally & ethically,
We must keep trying
To change!)
Exude respect!!!
Don't slime into the swamp of Trumpian/Jacksonian
        bullying,
Trumpian abuse,
Trumpian arrogance, and ignorance of ecology.
Avoid this vacuum of moral and ethical values.
Start by communicating softly and effectively;
Use the Lakota lingua,
A language void of harsh, cursing words.
Do not hit and hurt with words, limbs, arms or armaments,
No matter what language you speak.
Follow the Lakota Way.
Moreover, as Aunt Donette Lone Hill gently stresses
F-words should be "fun," "fantastic," "friendly," "fair," "faith-
ful,", "fearless," "flawless."
Never go over Grandpa Garland Not Afraid's red line
        example for
Displaying strong emphasis;
That is, "Dear one, would you kindly get the hell out of the
        way?"
Yes!!!
Communicate! Listen!
Empathize!
Try to walk miles in their sandals, moccasins, shoes, boots.

LISTEN!

Listen!
Two-year-old little Lyndon Gaddie
Cries out loudly
Because his limbs are asleep
From over an hour on uncle paul's shoulders
Without a squeak of a complaint.
His appendages are now driving him crazy.
He's pleading through his tears
For us to massage his numb and aching legs
Before he's stuffed into a cramping high-chair.
Listen in some manner
To all which is connected!
All lives have been difficult and tough ...
And different from those of others!
Recognize the pain and suffering in all,
And help get feelings out in the open,
Out on the table,
In order for them to be dealt with in solidarity,
Collectively
And softly,
As a whole and holistically empathetic
And communicating community.

Listen!
Steadfastness, hard-headedness, stubbornness are good.
We've got to persevere and be principled
But only up to a point.
So listen!
We also must be pragmatic and compromising.
Husband paul and wife Betsy,
And all you other partners,
YOU may be right.
But so may he or she.
The use of hard and abusive words,

To stand up for principles and your way
Are wasteful of life-changing power
And destroy quality life on both sides
In both life systems.

Listen.
Camping, roughing it in a cold and hot tent,
Enduring mosquitoes, deer flies, ticks, and other critters
Are not as pleasurable to some as to others.
Listen. Communicate.
Help each other bear the pains,
But, moreso
Help each other to enjoy the pleasures of Nature,
The joyful experience of the raw Mother Earth.

Listen!
Mentors Jeffrey and Austin have wisely articulated,
In the process of learning about quality life for all
For as long as possible,
That we must have a collaborative system.
We should respect schedules and agreed-upon
responsibilities.
However, collaboration also means flexibility through loving
communication.
Listen to the plants, the photosynthesizers.
They are for the most part
The only real producers!
Beneficial, nourishing, healthy medicinal plants.
And, to communicate about them we need names,
Lakota names,
Occupier-gringo names,
Universal, Latin names.
Listen to the aquatic organisms.
Plenty of pollution-sensitive biota?
Then you've got indications of
A healthy water system,

A healthy watershed,
A healthy ecoregion.
Listen to the biodiversity.
You have plenty of plant and insect and other biota
diversity?
Then you probably have a healthy ecosystem,
A healthy Earth,
A healthy oikos.

Listen.
Chemtrails? Contrails? Vapor trails?
Whatever.
They are from very artificial, life-draining systems.
And, yes, they are often produced by the military, by the
government.
"Our" occupying government.
"Our" government.
By "us"!

Listen!
To history of a culture and its current status:
Paleolithic-type hunter gatherers,
Then mostly agriculturists,
Then hunter-gatherers on horses for 200+ years,
And now we are all de facto globalized and agro-industrial
dependent.

Listen.
Tipis. Tents. Houses. Fences. Borders. Boundaries.
They are of importance
For shelter, storage, and protection.
Still we must readily share food, fiber, shelter, territory.
We have too many barriers,
Too many walls.
Do some lying out in the Natural, in Nature.
Let the mosquitos, deer flies, and ticks draw blood.

Become a part of dynamic homeostatic symbioses (Nature).

Listen!
Expansive, powerful, energy draining missile systems.
Big expensive, life-destructive military-government/
      bureaucratic-industrial.
Is it good to have such power?
Could be better utilized for learning;
For ecology across campuses and curricula of all human
      organizational entities;
For Generations Indigenous Ways all over the world?

Finally ... Listen! ¡*Escucha*!
Nightingale, cook, future mentor
Jamie Milapashne is singing The Song.
She beautifully wraps this poem up with
"We ALL have to be lovingly connected,
Because we are necessarily connected!"

....

An Indigenous Poem (Real Poetry!)

"Earth, Teach Me
Earth teach me quiet ~ as the grasses are still with new light.
Earth teach me suffering ~ as old stones suffer with memory.
Earth teach me humility ~ as blossoms are humble with
      beginning.
Earth teach me caring ~ as mothers nurture their young.
Earth teach me courage ~ as the tree that stands alone.
Earth teach me limitation ~ as the ant that crawls on the
      ground.
Earth teach me freedom ~ as the eagle that soars in the sky.
Earth teach me acceptance ~ as the leaves that die each fall.
Earth teach me renewal ~ as the seed that rises in the spring.

Earth teach me to forget myself ~ as melted snow forgets
    its life.
Earth teach me to remember kindness ~ as dry fields weep
    with rain."

http://www.sapphyr.net/natam/quotes-nativeamerican.htm

....

Waunspe
Wacekiya

Let us learn on our own
And let us prod ourselves
to actively do so.
True learning has to be hands on,
Often painful.
We've got to get dirty.
But it is worth it
For all of life ...
For dynamic homeostatic symbioses.

Camarón que se duerme
se lo lleva la corriente.
The shrimp that falls asleep
is carried by the current.

(There were many Wise words
spoken by the peoples
of Generations Indigenous Ways
In Yellow Bear Canyon
On a fatigued but powerful
Thursday, June 27, 2019. ...
I learn most about sustainable social fabric
And sustaining ecological systems
From trips to Honduras, Nicaragua, Mexico, and Pine Ridge
    and other indigenous reservations;
I learn more than from any reading I do,
Or from living elsewhere.)

Sueños espirituales

San Pedro Sula, June 3, 2019

I have many dreams, I think,
Especially while I am awake.[1]
Perhaps that is all I do…
Dream!
It aggravates others,
But it doesn't bother me much.

Last night it was of the sparkling, dancing eyes
Of Esmeralda Santos Ramirez,
The mother of Yony and Rafael Martinez,
Grandmother of *mi compadre* Luis Romero's wife Laira …
In Zapote, Comayagua, Honduras …
Earth.
A vision of a small diversified farm,
Struggling for sustainability
In the natural resource rich mountainous area around
      Siguatepeque.
Of a family that is glued together
After generations, varying economic status and power,
And political changes.

Because of Esmeralda's spirit
My life was especially enriched.
If only women like her
And Mom Louise
And Berta Isabel Cáceres Flores,

---

1. *https://marikasculptures.com/2011/03/15/creativity-sustainability-at-siempre-sustainable-network/*

Ellen Johnson Sirleaf, Michelle Obama,
Winona LaDuke, Ruth Bader Ginsburg,
And, yes, Alexandria Ocasio-Cortez and Elizabeth Warren,
(Maybe Nikki Haley?)
Ran the world.

Not Donald Trump!!!
Not many of the male Democratic candidates
For president of the *EEUU*.
Not Margaret Thatcher,
Or Michele Bachmann, Kellyanne Conway, Sarah Huckabee
        Sanders,
Or Teresa May.
Or Eva Peron.

Good women, saints,
Practicing getting along with all,
Idealistically and pragmatically,
Locally and across the globe
With healthy doses of the caution and tentativeness.
Developing systems of living the 7 S's ...
Sabiamente, Simply, Smally, Slowly, Steadfastly, Sharingly,
        Sustainably.
Developing a process of VV->^^.
Strengthening social fabrics
And dynamic sustainable ecological systems.

Practicing intolerance!... intolerance!!...
Intolerance!!! of War.
Tearing down walls.
Refusing to fund militaries, arms, armaments
At the expense of holistic ecological education
And robust, profound, and comprehensive health care
For all.

Creating *pan de elote*
For family, friends,
Welcoming all of the Earth.
Sharing *sandía.*
Holistically embodying
And manifesting
Biophilia,
Love for the *naturaleza*
Including all humans
Of all colors, sexualities, ethnicities, and tribes.

II.   LIMITS
(Carrying Capacity and Biocapacity)

You are squeezing US out!

STOP!

We have Overshot the Earth's Carrying Capacity.
Humanity's Ecological Footprint has <u>overshot</u>
the limits of biocapacity or sustainability.

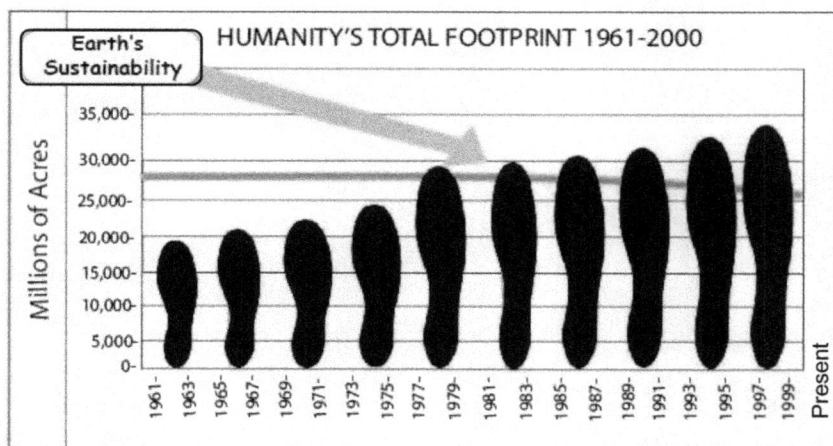

Earth's Sustainability

HUMANITY'S TOTAL FOOTPRINT 1961-2000

Millions of Acres

35,000-
30,000-
25,000-
20,000-
15,000-
10,000-
5,000-
0-

1961- 1963- 1965- 1967- 1969- 1971- 1973- 1975- 1977- 1979- 1981- 1983- 1985- 1987- 1989- 1991- 1993- 1995- 1997- 1999- Present

Modified from Footprintnetwork.org

¡Ya basta!

When is enough ... enough!??!
¿Para mí?
¿Para la familia?
For my tribe?
For my locale?
For a Trumpian EEUU.
For the bloated Haves on this Eaarth.

Why? There are limits. (And we are over carrying capacity as well as destroying biocapacity!)

Please leave some Earth for me!!

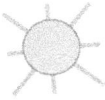

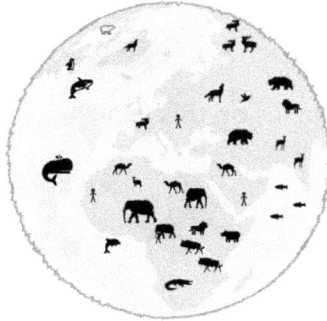

EARTH
50,000 BCE

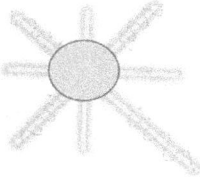

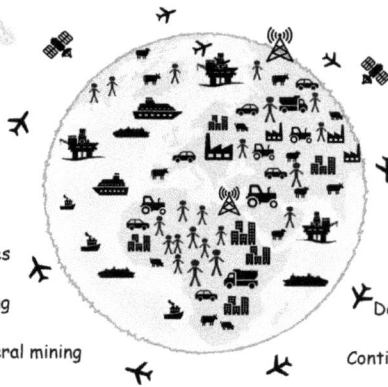

EAARTH
2020+

Cars

Big Cities

Asphalt-, concrete-cover

e-, internet-activity

Domesticated Animals

Continuous Grazing

Melting ice, rising sea levels

Acid estuaries

Clear-cutting

Big Homes

Industry

Plowing

Phosphate, other mineral mining

CO_2

Etc., etc.

Why? There are natural resource limits. We have to "draw a line in the sand" with respect to per capita energy transformation (energy "used") and ecological footprints.

> Haven't you guys heard about biological carrying capacity? Your greed and reckless /conspicuous consumption are making us history!

Consumption by the Powerful Over Time

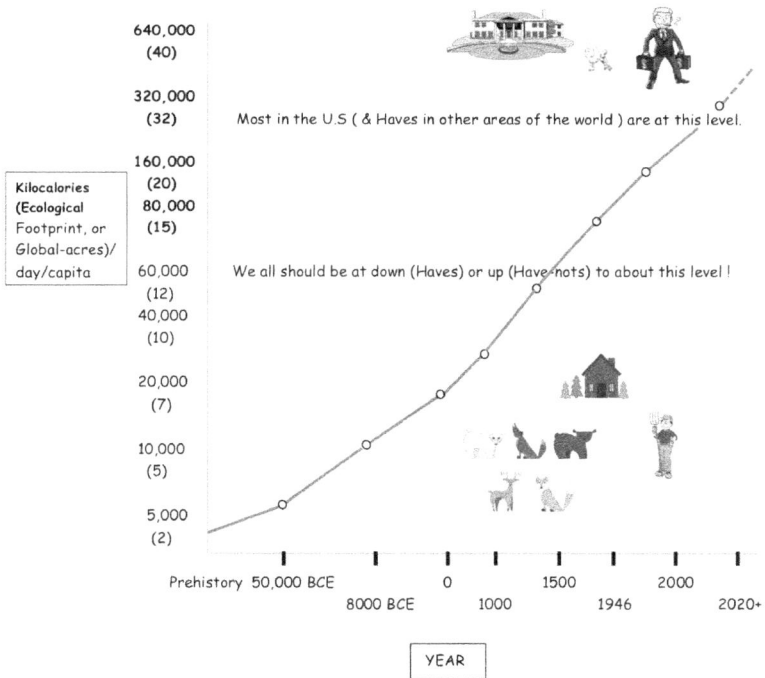

Kilocalories (Ecological Footprint, or Global-acres)/ day/capita

640,000 (40)

320,000 (32) — Most in the U.S (& Haves in other areas of the world) are at this level.

160,000 (20)
80,000 (15)

60,000 (12) — We all should be at down (Haves) or up (Have-nots) to about this level !
40,000 (10)

20,000 (7)

10,000 (5)

5,000 (2)

Prehistory 50,000 BCE   0   1500   2000
8000 BCE   1000   1946   2020+

YEAR

To Kill! or Not to Kill?

"To live, we must daily break the body and shed the blood of Creation. When we do this knowingly, lovingly, skillfully, reverently, it is a sacrament. When we do it ignorantly, greedily, clumsily, destructively, it is a desecration. In such desecration we condemn ourselves to spiritual and moral loneliness, and others to want." Wendell Berry (2003)

While enjoying tennis matches with seven gracious women on one cool (70 degrees F) fall morning at Starcke Park here adjacent to the Guadalupe River in Seguin, Texas, a cute little fall webworm crawled across the court. Like my attitude toward ants and spiders and most life forms, I felt admiration and respect for this little fuzzy critter and took care to let it continue in its seemingly focused journey toward further enjoyment of life.

Nevertheless, I am not reluctant to kill a "lowly" cockroach. Having been conditioned to immediately think of cockroaches as despicable, I will not rest until they are more or less eradicated from my home when detecting one or more of these creatures. Moreover, this tough attitude toward blattids doesn't stop in my abode! In Seguin, Texas there is a restaurant which has an excellent reputation for good tacos. I currently do not give this place my business after seeing a German cockroach on its vending-counter some five-plus years ago.

The only other organisms for which I have such a "genocidal-attitude" are rats and mice. My disdain for these creatures is not unlike that toward cockroaches; I absolutely do not want these creatures moving into my household or attached environment of living and working.

Other killings for which I feel truly little guilt are those committed against screwworms (New World species) early in my life. Perhaps, I have lived a somewhat sheltered life, but some of the most gruesome experiences in my 74-plus years were screwworm-infested cattle heads, calf navels, and pig scrotums. As a young side-kick of and laborer for my father in the 1950's and 60's, I helped him doctor, or treated myself, many wounds in hogs and cattle which were infested with this flesh-eating insect parasite with the black Smear-62 screw-worm killer or EQ-335 (or rarely an aerosol with the chlorinated hydrocarbon, lindane, as an active ingredient).

During my freshman year (1960-61) at Devine High School in vocational agriculture, I participated in a Future Farmers of America-Greenhand Farm Radio contest in which our topic was the proposed screwworm-eradication program for the southwestern U.S., or basically Texas. We discussed the biology of the screwworm, the promising sterile insect technique for eradication of this insect, and a proposed beef cattle check-off program to fund the eradication program.

By the time of my graduation from high school in 1964 the screwworm was declared eradicated from Texas. During my high school years Dr. E. F. Knipling's eradication program using airplane releases of irradiated and sterilized screwworms, which had been raised at an old Air Base in Mission, Texas, had been put into high gear and was successful. (There was a resurgence of screwworms in the 1970's reported to me from back home by my brothers while I was working on my doctorate in entomology at the University of Florida. I was subsequently fortunate to be able to attend a fabulously informative international conference in Gainesville, FL, addressing the dynamics of field populations and challenges which had arisen with genetic changes [micro-evolution], and need for better quality control of lab-reared screwworms. We eventually did eradicate screwworms from North and Central America but still have screwworms in South America, and there have also been recent outbreaks of New World screwworms in North Africa and the Florida Keys. The outbreaks in Africa and the Keys have been subsequently snuffed out!)

Now there have also been concerted efforts in the 1950's and 60's and even up until current times to eradicate species like screwworms, which were considered to be immensely pestiferous. And they were done without much adherence to the Precautionary Principle. Moreover, in their 1998 review of eradication and pest management, J.D. Myers et al. stated, "Cost-benefit analyses of eradication programs involve biases that tend to underestimate the costs and overestimate the benefits.[1]"

---

1. Myers, J.H. et al. 1998 Eradication and pest management. Annual Rev. of Entomology. 43. 471-91

Anyway, I will emphasize that the screwworm eradication effort was steamrolled through by E. F. Knipling in his powerful leadership role in USDA entomological research. It did employ many of my colleagues in entomology from the southwestern U.S. and was a successful program in the southeastern and southwestern U.S., Mexico, and Central America. On a very personal note, in 1980, as a young research scientist and coordinator of a fall armyworm conference, I had the pleasure of being invited to share some Wild Turkey whiskey with the stately and venerated and very knowledgeable and kind, Dr. Knipling in his hotel room in Biloxi, MS, and of discussing, one on one, his proposed talk on area-wide strategies of integrated pest management, including biological control. … Dr. Knipling, who died at 90 years of age, was 71 or about my current age at the time.

To reiterate and conclude my screwworm part of the story, in summary, the screwworm, a mostly tropical species, was relatively easy to eradicate from North America, and the unintended detrimental ecological consequences were probably relatively minor because of the generally low densities and relatively small biomass of the species. Therefore, as mentioned in the previous paragraph, we did not ponder too long upon the Precautionary Principle before deciding to eradicate the screwworm. (I must mention that removal of screwworms did have some unintended consequences and surely did affect populations of white-tailed deer and other wildlife species and populations.)

Another species with which we have been successful in eradication efforts is the cotton boll weevil. We are close to eradicating this late 19th century/early 20th century invader species from most cotton-growing areas of the United States. And, even though eradication strategies have employed the use of widespread applications of biocides along with other tactics, for the long-term there will be a probable dramatic net reduction in the use of biocides for cotton production in the U.S. The boll weevil is a key pest of cotton, and when broad-spectrum biocides have been used for its control, the detrimental impact on parasitoids and predators of the cotton bollworm and other insects has resulted in a number of secondary pests, including the cotton bollworm.

Decisions to eradicate some invasive species (but generally not "invasive" agricultural crop-species and domesticated animals), resulting from the Columbian Exchange, which University of Texas' esteemed professor, Alfred Crosby, labeled "Ecological Imperialism," seem to be even easier to make than the examples chosen to be presented herein in my introduction to this piece because of my own lifetime experiences. This is especially so for exotic invaders of islands.

However, the situation is quite different for many species, e.g., when contemplating eradication of a mosquito species or even some particular populations of these pesky piercing-sucking feeders. Despite the serious health challenges caused for humans by mosquitoes, the huge numbers and biomass of mosquitos in symbioses[2] (i.e., "nature"), and their role as key components of important food webs and biodiversity cause us to think at least twice when we ponder eradication of even *Aedes aegypti,* vector of so many human diseases, or *Anopheles* mosquitoes, vectors of dreaded malaria.

(I do wish to insert at this point, that I am in accordance with the famous sea turtle biologist, Archie Carr, who taught me much about community ecology, in stressing that we should for the most part leave snakes, including rattlesnakes, and spiders, scorpions, wasps, and bees, etc. alone. Generally, let them live. Let them be!)

Shifting to another level or aspect of killing, as is the case with most humans I have slaughtered many plants for various reasons over my lifetime. However, because we generally share less than 60% of our respective genes, we are not empathic to their feelings and consciousness. Moreover, because the eating of them and other primary producers are what we depend upon, directly or indirectly, for the matter and energy resulting in human life, this slaughter of plant lives does not cause much distress for us. We will swear off eating of our dear animal kin but never swear off killing and ingesting ALL sacred life forms ... including some dear animal kin, dear plant kin, AND dear loveable ones of other phyla. (Obviously, deciding to abstain from killing and eating all the various wonderful life forms would be to commit suicide.)

A sobering fact is, however, that the heavy load on the ecosphere of humans and their domesticated species is of itself an effective killer of "climax" vegetation and communities and symbioses, or "nature," to the extent that current numbers of humans and domesticated species should be considered to be immoral and unethical. On the other hand,

2. I lose some readers and risk a reduction in communicating when using some of these specialized ecological terms. Nevertheless, I feel strongly that we need these terms in our lexicon to communicate, develop, and grow toward consilience and empathy and begin to collectively develop sustainable community.

as one little example for making a world better with current numbers of humans and cattle, proper rangeland management and the utilization of grass-fed domesticated bovines, browsers, and free-range animals as food is much more appropriate than feedlot-confined or caged domesticated animals with respect to killing of individuals of various species, depletion of biodiversity, and extinctions as well as aspects of humaneness. Maintenance of rangeland systems generally results in much less destruction and chaos in symbioses than plowing these systems out for wheat and other small grain production systems or even introduced exotic pasture grasses. These more natural rangeland systems of food production result in the killing of less soil biota and other biota and the realization of much more biodiversity–than grain, vegetable and most fruit and nut production systems. (By the way, the Land Institute, Salina, Kansas is working steadily and vigorously to improve current monoculture-systems of grain with breeding and development of more ecologically appropriate perennial grain-, oil-, and legume-crops, which might be utilized in more diverse intercropping systems involving less tillage and less synthetic fertilizers and biocides. *https://landinstitute.org/about-us/*)

We are going to have to continue to kill to healthily survive as human individuals, demes, and populations and to survive as a species. But where do we draw the line of killing in these and other aspects of living/killing systems? When we attempt to put all into perspective, it is generally not an easy task to make these killing decisions. On the other hand, in another arena of killing, war and capital punishment, for example, are de facto premeditated acts of murder of living lived lives and they definitely have no moral or ethical currency.

Decisions concerning conventional birth control are not so clear-cut. Human birth control can be effective in reducing growth in populations of *Homo sapiens*, which generally results in more resources and habitat for all others. When one individual, deme, or population dominates an inordinate amount of energy, matter, food, or habitat, this action inevitably results in killing of other lifeforms, including other humans, or the prevention of adequate matter and appropriate energy transformation for future living beings. But depending on the type of birth control, it does also result in the killing of a partial human, i.e., a sperm or ovum, or of a developing human zygote, embryo, or even a fetus. Moreover, in addition to many other aspects of a dynamic biosphere to consider, living systems with an abundance, or even an overabundance of humans can provide resources and excellent habitat for some species.

124

The somewhat arbitrary[3] lines in the sand to effectively reduce killings, which I posit herein and are presented in various illustrations in this little book on applied ecology, can lighten the ecological footprint and reduce embodied human-appropriation of net primary productivity. Moreover, the immensely powerful must share that power with the relatively powerless, poor, and disenfranchised. We do need to realize per capita equity of <70,000 kilocalories/day and perhaps ca. <$50,000/year in 2020 U.S. dollars.

Finally, increase in unsustainable jobs, GDP, the power of the military-industrial complex, human and domesticated animal numbers, and the corresponding rampant human development and artificialization of what was a robust and healthy symbioses are indicators of and processes of a global killing-machine.[4] To slow down this killing[5] we desperately do need to live by the mantra of "Sabio, Simple, Small, Slow, Steadfast, Sharing, SUSTAINABLE."

(Perhaps readers of this piece will begin to understand why I had feeling of total sickness upon the election of Donald Trump and why I have generally felt sicker ever since. Trump and his policies are the antithesis of a process of living in solidarity, smartly and sharingly, toward a healthier symbioses and a healthier humanity with less killing.)

In slightly paraphrasing Wendell Berry, when we break the body and shed the blood of creation knowingly, lovingly, skillfully, reverently, it is a sacrament. When we do it ignorantly, greedily, clumsily, destructively, it is a desecration.

---

3. The <70,000 kilocalories/capita/day originally came from a napkin calculation to keep a world of 9-11 billion at a similar energetics level to, or just below, what it is today with ca. 8 billion humans.

4. There are numerous systems to realize indicators of sustainability which have been proposed and researched. Five indicators that might rise to the top are: the Gini coefficient, embodied human appropriated net primary productivity, ecological footprints and biocapacities, embodied kilocalories "used"/per capita/day, and soil sealing per capita. However, there are many others of importance.

5. Recent scientific reports indicate global insect numbers are falling dramatically on this Anthropocene Eaarth. As an example, a study in 2017 "showed a 76 percent decrease in flying insects in the past few decades in German nature preserves. " SA Express-News October 16, 2018. *https://www.washingtonpost.com/news/speaking-of-science/wp/2017/10/18/this-is-very-alarming-flying-insects-vanish-from-nature-preserves/?utm_term=.347d50c3395*

Why? "Artificialization" takes from Nature. Nature is natural and "perpetual!" Humans are dependent on Nature. Why would we continue to replace Nature???

Artificialization is being human! (But you *huumaaans* have really gotten carried away!!!)

Artificialization Is Increasing While ... Nature, The Natural Resource Base, The Foundation for Quality Life, And Biocapacity Are Diminishing.

2020 +

Why? Dirtying our nest!

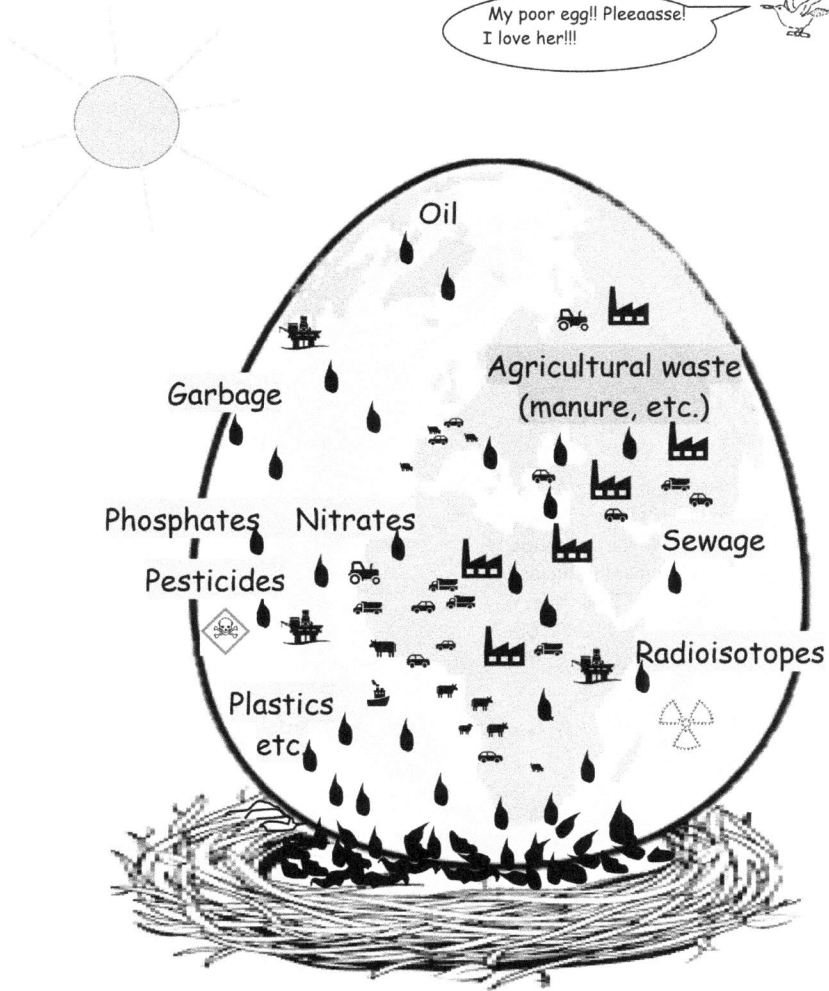

Why? Negative externalities (Generally, negative consequences affecting others, including future humans, without being reflected in the current cost of the goods or services.)

LASTING GIFTS OF BLIGHT FROM TRANSNATIONAL CORPORATIONS
(and THE HUMANS OF THESE CORPORATIONS):

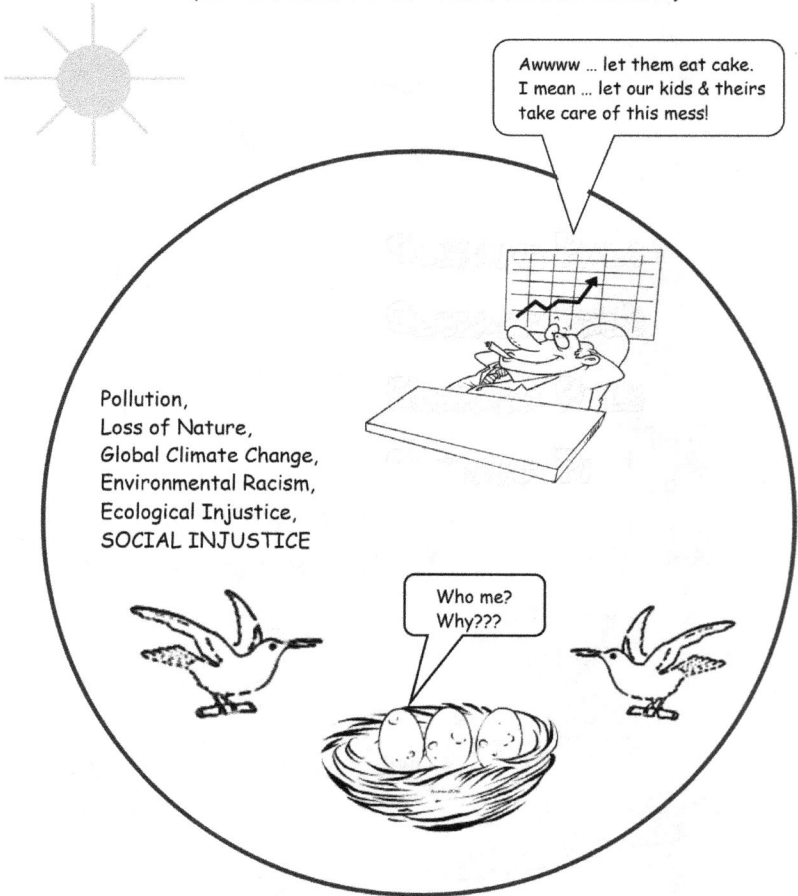

Awwww ... let them eat cake. I mean ... let our kids & theirs take care of this mess!

Pollution,
Loss of Nature,
Global Climate Change,
Environmental Racism,
Ecological Injustice,
SOCIAL INJUSTICE

Who me?
Why???

Why? Current agriculture systems are not sustainable!

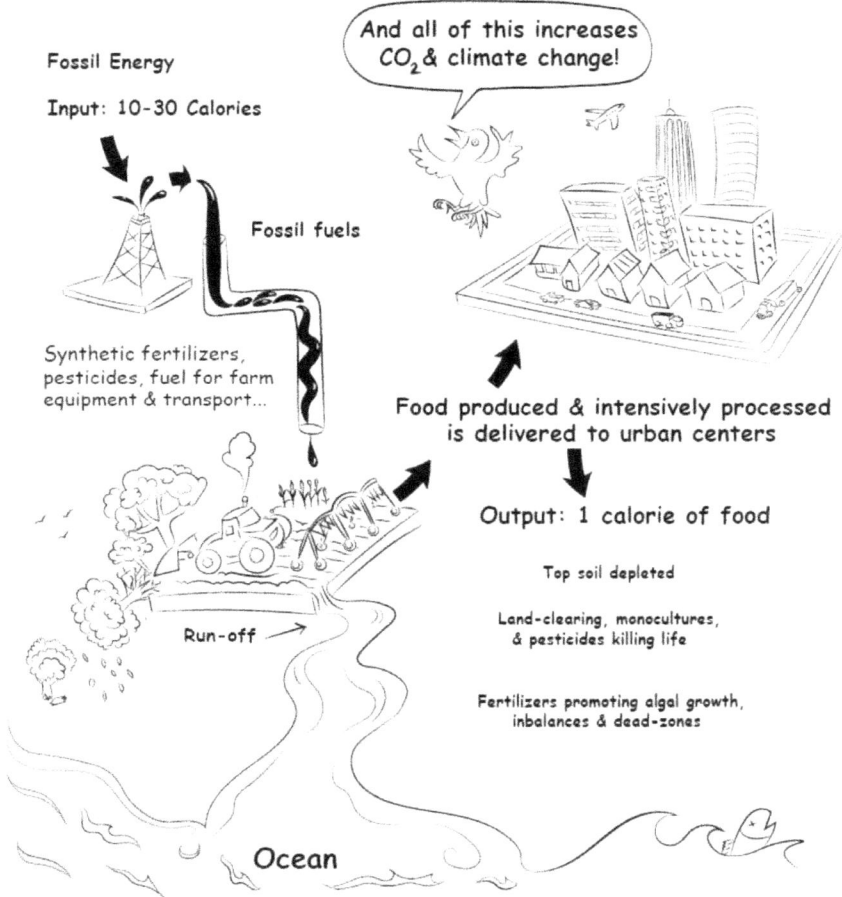

And all of this increases CO_2 & climate change!

Fossil Energy

Input: 10-30 Calories

Fossil fuels

Synthetic fertilizers, pesticides, fuel for farm equipment & transport...

Food produced & intensively processed is delivered to urban centers

Run-off →

Output: 1 calorie of food

Top soil depleted

Land-clearing, monocultures, & pesticides killing life

Fertilizers promoting algal growth, inbalances & dead-zones

Ocean

Why? High input/throughput conventional agriculture is
dependent on disruptive fossil energy inputs.

Haber-Bosch is a
major culprit!

High Input/Throughput
Conventional Industrial Agriculture

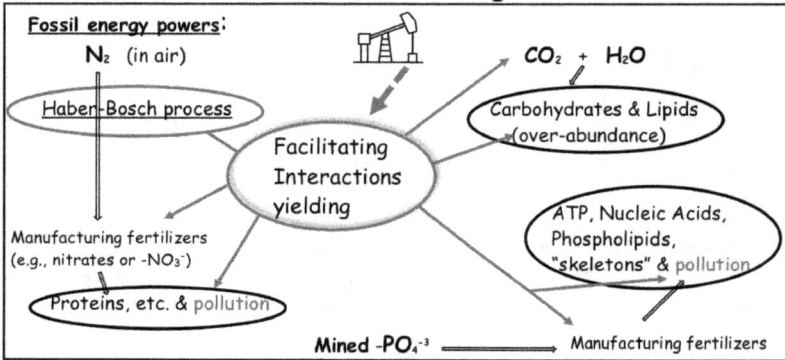

Fossil energy powers:

N_2 (in air)

CO_2 + H_2O

Haber-Bosch process

Carbohydrates & Lipids
(over-abundance)

Facilitating
Interactions
yielding

ATP, Nucleic Acids,
Phospholipids,
"skeletons" & pollution

Manufacturing fertilizers
(e.g., nitrates or $-NO_3^-$)

Proteins, etc. & pollution

Mined $-PO_4^{-3}$ ————————→ Manufacturing fertilizers

Low Input/Throughput
Sustainable Agriculture

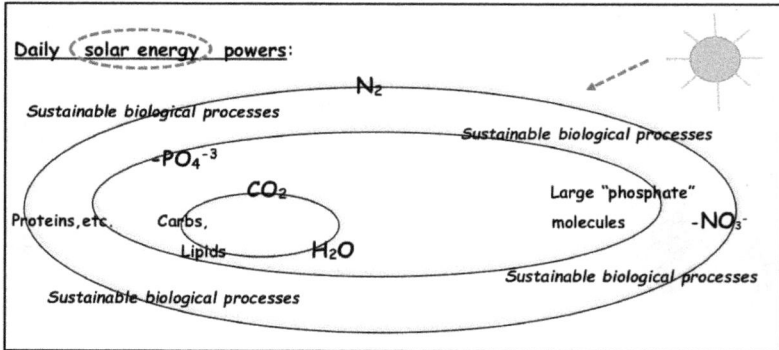

Daily (solar energy) powers:

N_2

Sustainable biological processes

Sustainable biological processes

$-PO_4^{-3}$

CO_2

Proteins, etc.   Carbs,   Lipids H_2O

Large "phosphate"
molecules $-NO_3^-$

Sustainable biological processes

Sustainable biological processes

What? Focusing & "picking on" golf courses, golf course managers, and golfers (**a microscosm of Anthropocene Eaarth**). (Of course we could also easily pick on many other professional & highly artificial sports, fitness, and recreational endeavors, including college football and treadmills, etc., etc., etc..)

Conventional golf courses aren't really all that much fun, especially for us wild ones! They are very artificial, lack biodiversity, don't rely on native that much, and use considerable amounts of pesticides and synthetic chemicals.

A sustainable golf course would be of locally-adapted native plants, would use no pesticides and synthetic fertilizers, would be maintained with grazers & browsers, and would mostly be walked.

III. Ecological Principles and Processes and Appropriate Values

Ecological Principles and Processes and Appropriate Values

(Permaculture uses a flower to illustrate ethics and principles focused initially in the critical domain of Commons, Land, and Symbioses enhancement. Permaculture then evolves by progressive application of principles to the integration of all seven domains.)

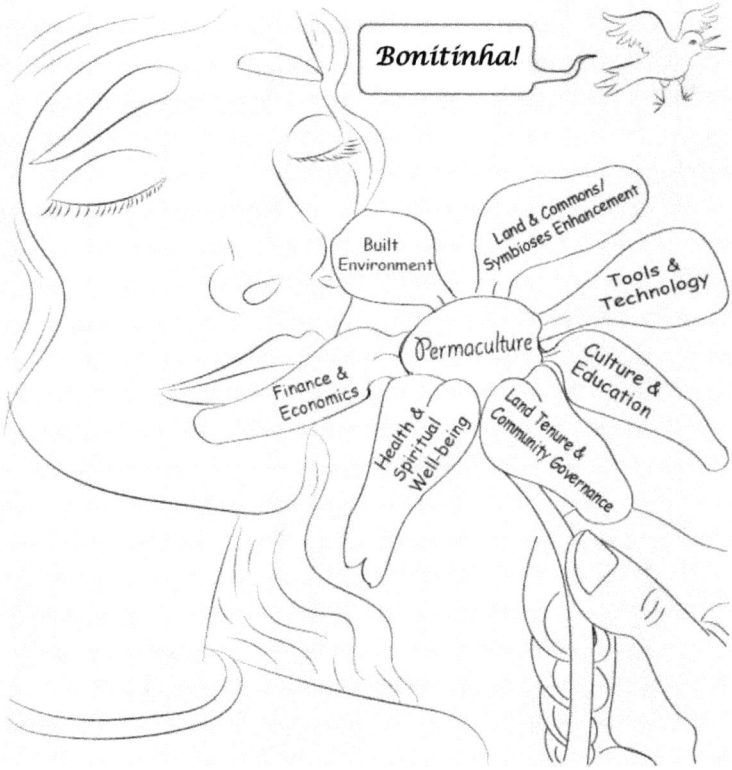

How? All education is ecological education!
(Paraphrased from David Orr.)

You just don't know me and how important we are to each other.

Ignorance is the major barrier to sustainability/
"Positively Ethical Applied Community Ecology-PEACE."

What there is to know *(and it's actually much, much more than depicted on this page)* : ???????????????????????? ??? ???????????????????????????????

Versus

What we know: Eureka ! ! ! !

We'll always be ignorant. However, we can begin to learn
more effectively, significantly increase ecological
knowledge, and do a better job of thinking critically.

1. Ecological principles
   and processes

2. BIOPHILIA

As we develop, knowledge of " 1 "
and hone our good values in " 2 "
we can begin thinking critically and
functioning morally and ethically in a
socially just, humane, and ecologically
sane manner.

Principles for Understanding and Sustaining the Earth

(Miller 1990)[1]

Recycling mineral resources takes energy, which in being produced and used, causes pollution and environmental degradation (recycling-is-not-the-ultimate-answer principle).

To reduce resource waste and resource supply interruptions, get as much as possible of what we need locally, and dispose of or recycle wastes locally (principle of localism).

Stress the use of perpetual and renewable resources and use renewable resources no faster than they are replenished by natural processes (principle of sustainable yield).

Everyone is downwind or downstream from everybody (principle of the global commons).

Organized or concentrated energy is high-quality energy that can be used to do things; disorganized or dilute energy is low-quality energy that is not very useful (principle of energy quality).

In any conversion of energy from one form to another, high-quality, useful energy is always degraded to lower-quality, less useful energy that can't be recycled to give high-quality energy; we can't break even in terms of energy quality (second law of energy or law of energy-quality degradation).

Everything runs on moderate- to high-quality energy, that cannot be recycled, so choose and use energy resources wisely (principle of energy use and flow).

Everything is connected to and intermingled with everything else;

1. From a longer list in: Miller, G. Tyler, Jr. 1990 Resource Conservation and Management. Wadsworth Publishing Co., Belmont, CA

we are all in it together (second law of ecology, or principle of interrelatedness).

The earth's life support systems can take much of stress and abuse, but there are limits (laws of limits).

No population can keep growing indefinitely (principle of carrying capacity).

Nature is not only more complex than we think but more complex than we can ever think (principle of complexity).

The market price of anything should include all present and future cost of any pollution, environmental degradation, or other harmful effects passed on to society and the environment (principle of internalizing all external costs).

The more things you own, the more you are owned by things (principle of overconsumption and thing tyranny).

Human population growth ultimately makes democracy and individualism impossible (principle of freedom erosion).

Do not ever call yourself a conservative unless what you want to conserve is the earth (principle of true conservatism).

We are a part of nature (principle of oneness).

We are a valuable species but are not superior to other species; all living beings, human and nonhuman, have the same inherent worth (principle of humility).

Every living species has a right to live, or at least struggle to live, simply because it exists; this right is not dependent on its actual or potential use to us (respect-for-nature principle).

Our role is to understand and work with the rest of nature, not conquer it (principle of cooperation).

It is wrong to treat people and other living things primarily as factors of production, whose value is expressed only in economic terms (economics-is-not-everything principle).

We have a right to protect ourselves against harmful and dangerous organisms but only when we cannot avoid being exposed to such organisms or safely escape from the situation; in protecting ourselves we should do the least possible harm to such organisms (principle of self-defense).

We have a right to kill other organisms to provide enough food for our survival and good health and to meet other basic survival and health needs, but we do not have such rights to meet nonbasic or frivolous wants (principle of survival).

When we alter nature to meet what we consider to be basic or nonbasic needs, we should choose the method that does the least possible harm to other living things; in minimizing harm it is in general worse to harm a species that an individual organism, and still worse to harm a biotic community (principle of minimum wrong).

We must leave the earth in as good a shape as we found it, if not better (rights-of-the-unborn principle).

No individual, corporation, or nation has a right to an ever-increasing share of the earth's finite resources; do not let need slide into greed (principle of enoughness).

To prevent excessive death of people and other species, people must prevent excessive births (birth-control-is-better-than-death-control principle).

Everything we are and have or will have ultimately comes from the sun and the earth; the earth can get along without us, but we can't get along without the earth; an exhausted earth is an exhausted economy (respect-your-roots or earth—first principle).

To love, cherish, and understand the earth and yourself, take time to experience and sense the air, water, soil, plants, animals, bacteria, and other parts of the earth directly; learning about the earth indirectly from books, TV, images, and ideas is not enough (direct-experience-is the-best-teacher principle).

Learn about and love your local environment and live gently within that place; walk lightly on the earth (love-your-neighbor principle).

How? Ecological literacy and ethics (ecological mindedness) across curricula (including continuing educations) and campuses of all human organizational entities

> 1.Humans & human entities need to understand **Why**? we need sustainable local/global community. 2.Generally the **What's**?, or day-in-and-day-out actions needed for lowering our ecological footprints, are known. 3.Then through ecology across curricula & campuses, etc., humans can address the **How**?

1. Realize regular intra- and inter-campus meetings to begin to come to a consensus of what Positively Ethical Applied Community Ecology, PEACE, is.

2. Each education/continuing education course is permeated with "Positively Ethical Applied Community Ecology/PEACE."

3. On and off campuses, "Positively Ethical Applied Community Ecology/PEACE" is always a work in progress.

4. There must be spiritual and scientific study along with community service toward regeneration and conservation of resilient, sustainable community.

Many, many small diverse campuses intimate with nature and with ...

Corporations   Homes   Government, bureaus   NGOs

Churches   Clubs, etc.

Businesses   Traditional schools

... continuing education over the lifetime of the student!

Why? Why do we basically refuse to realize "Positively Ethical Applied Community Ecology/PEACE" and resilient, sustainable (local/global) community?

How? Ecological literacy means having a generalist/liberal arts type of (continuing) education and learning as much as possible about the "Giants of Education," or those areas of learning which might be labeled as the most important.

> Yep! You are going to have to work at learning the basics if you are going to be a decent, moral, ethical citizen of the world, voter, critical thinker.

Using tools of education:
Communication (Language(s), MATHEMATICS, Computer Science)

> ¿ Te hablos matema'ticas ?

Playing important games:

ART Kinesiology/**SPORTS POLITICS** (Law, Military Science, ...)

> I vote for *futebol!*

Wrestling with the **WHY**?
Philosophy/Religion (Ethics, Morals, Values, Mores)

> Consilience

Understanding **What**? keeps living systems together at the basic foundation:
Physics (Energetics)

> Joules or $?

Roughly keeping tabs on artificialization:
Economics (Money, Marketing, Management, Accounting)

> Whether it is Joules or $, this is where it really counts!

Holistically Understanding Living Systems/Understanding it All:
ECOLOGY (Chemistry, Biology [Agriculture, Medicine, Dentistry]), Geography, Social Studies-**Anthropology**

What? Oxidation: *losing* electrons -- **Reduction:** *gaining* electrons
Chemistry is important to understand!

> Losing and gaining electrons is to a large extent what life and living systems are all about!

Redox Reactions in . . .

PHOTOSYNTHESIS

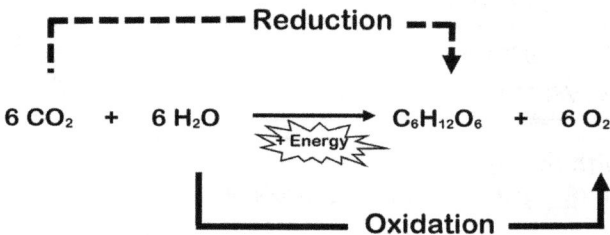

Reduction

$$6 CO_2 \quad + \quad 6 H_2O \quad \longrightarrow \quad C_6H_{12}O_6 \quad + \quad 6 O_2$$

+ Energy

Oxidation

C = Carbon   H = Hydrogen   O = Oxygen

AEROBIC CELLULAR RESPIRATION

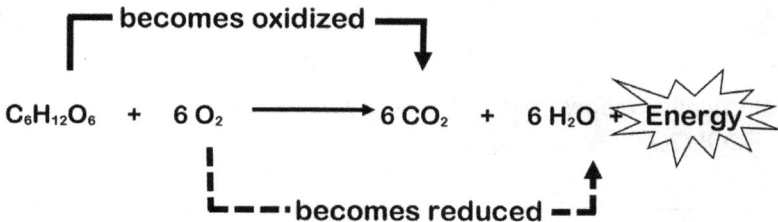

becomes oxidized

$$C_6H_{12}O_6 \quad + \quad 6 O_2 \quad \longrightarrow \quad 6 CO_2 \quad + \quad 6 H_2O \quad + \quad Energy$$

becomes reduced

The Second Law of Thermodynamics

(It Is in Many Ways the "Most Important" and Should Be "the First!")

"Gottamnit Leroy!" exclaimed Fritz very irately. "Ju may play goot pitch. But ju're a tamn disgrace ta da Jerman race!! I vent down to Haby's store and bought some of dat Alsatian sausage vit goot coriander for a nickel-a-link. Din I cuuked-tit-up and ate dat link sausage. Din I pooped tit out and put tit on my jard for da grass ta grow green. I can't believe ju Leroy—a Jerman—vasted dat poop like ju tid down da tamn poop-pot. You're a tamn disgrace ta da Jerman race!"

"Hell, Fritz!" exclaimed Bruno. " JU'RE a tamn tschame ta all us Jermans. I bought dat Haby link-sausage, cuuked tit up, ate tit, pooped tit out … and I put tit on my wegatable garten. I can't believe ju vasted dat sausage poop on dat tamn jard-grass!"

Karl jumped up off of his rickety old chair a-hollering. "I caan't believe ju tamn squandervers … all of ju! I'm tschamed to play pitch vit all of j'all. I vent down ta Haby's and got some of dat link sausage and cuuked tit. Din I cut dat casing and peeled tit off … weeery caaarefully. I ate dat goot link sausage and pooped tit out. Din I put dat poop in da casing and took tit down to Haby's and told dat Hans Haby, 'Hans, dis tamn sausage tastes like tschit! Give me my nickel back!' Hans took a bite and said 'Tamn!! You're right! Here's jur tamn nickel.'"

This story from the Alsatian-German-Texan land in which my wife Betsy grew up was my usual introduction to the Second Law of Thermodynamics for years in my principles of biology class at St. Philip's College, San Antonio, Texas. It was my attempt at a humorous story about how the Second Law of Thermodynamics was sort of miraculously circumvented by Karl. (I would follow it up in class by holding up some good candy like a pecan praline in my hand and ask if anyone wanted it. Of course, a bunch of these hungry students appreciative of good candy did want it badly, and they pleaded, "I do!!" in unison. Then I plopped it in my mouth, chewed it up and swallowed it, and teased them with, "You can have it after I'm through with it!")

The Second Law or Entropy Law was, I guess, briefly introduced to me in high school and college biology, physics and ecology. But it was economist Herman Daly, Daly's mentor Nicholas Georgescu-Roegen, and ecologist David Pimentel and energetics scientist H.T. Odum who really made me cogitate on it.

The Second Law states that in a thermodynamic process, the total entropy, or "disorder", of the participating systems increases. Also, as you transform energy, which cannot be created or destroyed, it tends toward uselessness. And one cannot recycle energy; recycling stuff always comes at a cost; perpetual motion is impossible; growth economics will eventually hit the wall (or I guess it already has!).

There are many other implications of the Second Law of Thermodynamics in positively ethical applied community ecology which should be considered. Hans Haby's nickel sausage was still useful as poop, but not as useful as it was in its store-bought form. And in using the sausage, each of the Alsatians did produce poop (or something which tasted like "tschit!").

Fossil energy resulting from the (inefficient) capture of solar energy through photosynthesis over millions of years can be very useful. However, rampant transformation and use of this or any energy source also results in pollution, stress and socio-political/economic (ecological) destruction and chaos. Moreover, BIG and Fast and Complex--whether it be houses, automobiles, geothermal air-conditioning, photovoltaics, windtricity, conventional or "organic" or sustainable agricultural food/fiber/shelter systems; cities, towns or villages; schools, businesses, governments, churches, or do-good non-profits—can be very problematic for whole systems. Order in one system creates chaos in other systems, communities, lives. Overdoing the built environment is detrimental to the natural resource base, biodiversity, and natural, efficient photosynthesis and biogeochemical and hydrological cycles. Too much Artificial destroys Nature.

"Small is Beautiful" (from E.F. Schumacher) and so are Slow and Simple. I'll state again my Wendell Berry mantra:

"To live, we must daily break the body and shed the blood of Creation. When we do this knowingly, lovingly, skillfully, reverently, it is a sacrament. When we do it ignorantly, greedily, clumsily, destructively, it is a desecration. In such desecration we condemn ourselves to spiritual and moral loneliness, and others to want."

Why? What? How? Ecological Principles, Processes, and Concepts

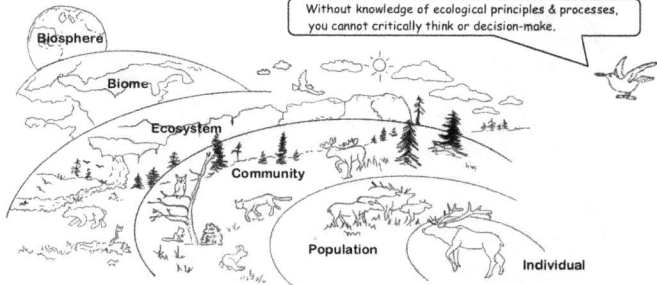

Without knowledge of ecological principles & processes, you cannot critically think or decision-make.

Biosphere
Biome
Ecosystem
Community
Population
Individual

- **Connectivity** and Hierarchy, Food Chains/Webs, Dominant/Keystone Species /Climax Species

- Biomes, Heat, Winds, Rainfall & Distribution, **Evapotranspiration Rates**

- **Energetics**; 1st & 2nd Laws of Thermodynamics; Energy Pyramid; Oxidation – Reduction, **Photosynthesis** /Respiration, Metabolism

- Ecosystem Blocks (1. **Soil & Air** 2. **Hydrological** 3. Dynamic **Biotic Communities** 4. **Solar Energy**), Biogeochemical Cycles

- Net Primary **Productivity**

- **Succession**

- Territoriality

Nature   Hands on Learning

Wisdom from the ages ⟹ Real Knowledge

Elders   Ecologists

- **Population Intra-actions** & **Dynamics**; r- & K-strategies

- Behavior & Interactions in Community—Communication (All the senses/ semiochemicals, predation/competition/symbiosis, interspecific)

- Regulatory Processes (Negative & Positive Feedback)

- Evolution

- Human: Giants of the education process, strategic planning, daily / capita energy transformation, Ecological Footprints, Human Appropriated Net Primary Productivity, other sustainability indicators

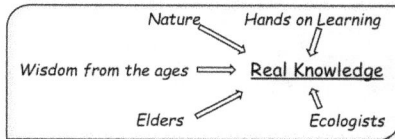

What? Energetics; 1st and 2nd Law of Thermodynamics; Energy
Pyramid; Photosynthesis/Respiration/Metabolism

Order (in one system) creates chaos
For another. Artificial order is
Chaos for Nature/Natural order.

1st Law of Thermodynamics—Energy cannot be created or destroyed and can
only be transformed.

2nd Law of Thermodynamics—When energy is transformed, disorder increases.
As energy is transformed, it tends toward uselessness. (Perpetual motion
machines and perpetual economic growth are impossible.)

Energy Pyramid

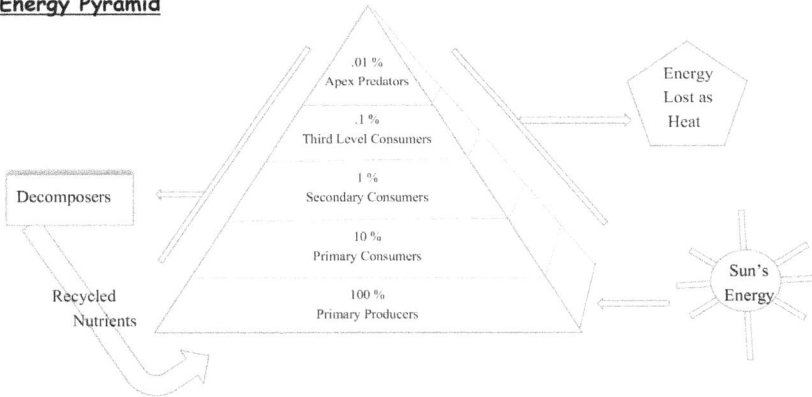

.01 %
Apex Predators

.1 %
Third Level Consumers

1 %
Secondary Consumers

10 %
Primary Consumers

100 %
Primary Producers

Energy
Lost as
Heat

Decomposers

Sun's
Energy

Recycled
Nutrients

Photosynthesis   **Respiration**

$6\ CO_2\ +\ 12\ H_2O\ +\ Light\ energy\ \longrightarrow\ C_6H_{12}O_6 + 6\ O_2 + 6\ H_2O$

Reverse the above reaction and energy is released, for e.g.,
<u>metabolisms</u> or the numerous regulated chemical
reactions of living-cells/-systems.

What? Biodiversity: Connectivity and Hierarchy, Food Chains and Webs, Dominant/Keystone/"Climax Species"

All life has meaning for YOU and for me!

Hierarchy of "building blocks" for organismal structure:

Quarks/Leptons—Protons/electrons/neutrons—Atoms/Molecules/Elements/Compounds—Molecules—Organic Molecules—Simple prokaryotic cells/Organelles—Eukaryotic organisms (Single celled/multicellular)—Tissue--Organs/organ systems—Multicellular Organisms—Demes—Populations—Communities—Ecosystems—Ecosphere—Solar system—Galaxies—Universes

An example of a food web involving cycling and recycling of above "building block" material through energy transformations:

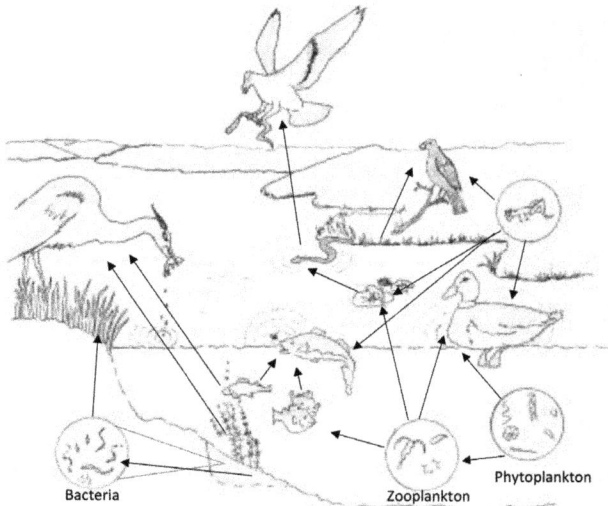

Phytoplankton

Bacteria

Zooplankton

Texas Wetland Community
Included in a food web may be **keystone, dominant, climax, increaser, decreaser,** and **invader** species.

What? Biomes. A grouping of the world's major ecosystems classified according to the predominent vegetation and characterized by adaptations of organisms to that particular environment. Climate is the major determiner.

Sunshine, winds, degree-days, rainfall, evapotranspiration, H_2O/land ratios, altitude, plant cover, CO_2/other air gases, etc. are keys to biome types.

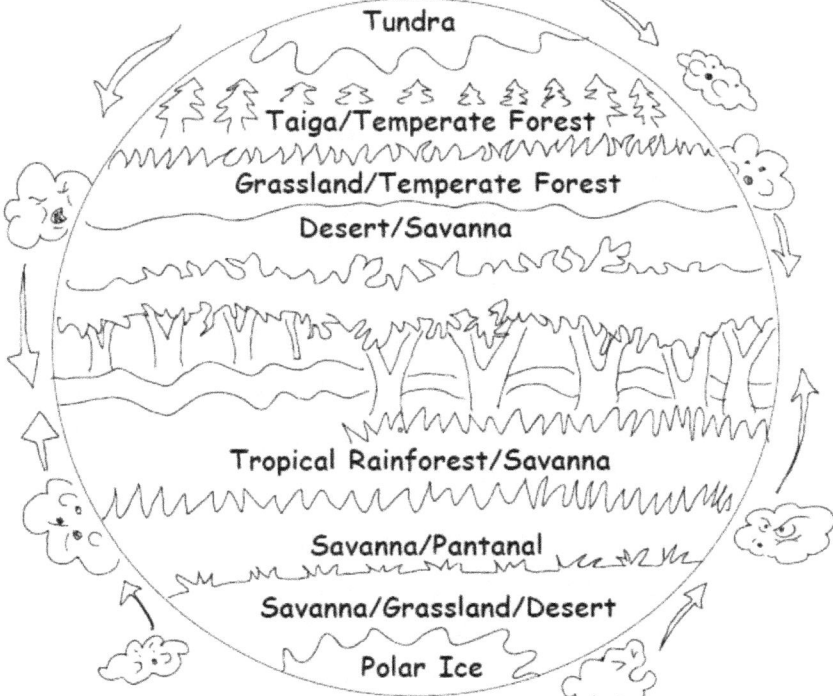

Tundra

Taiga/Temperate Forest

Grassland/Temperate Forest

Desert/Savanna

Tropical Rainforest/Savanna

Savanna/Pantanal

Savanna/Grassland/Desert

Polar Ice

Climatic Conditions for Diversity and Quality Life, including Humans

IV. Ecological Morals and Ethics: Especially a Profoundly Holistic Ethic of Reciprocity

An Appropriate Process. Nihilism?

Conception and life. … With sex. Without.
We begin. We sense. And we learn.
Still with so very much nil.
Fathomless mystery.
At times it frustrates.

But keep LIVING!
And learning.

Life's goal?

Care!!!

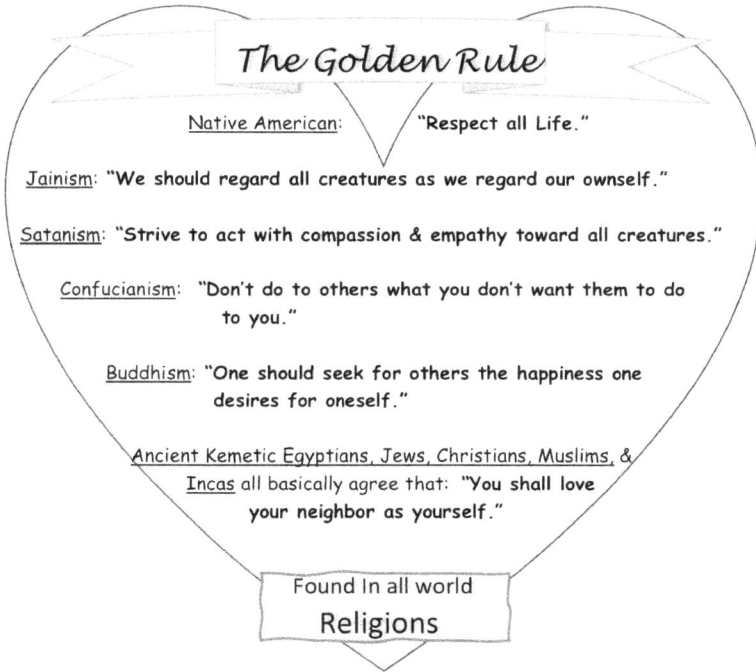

The Golden Rule

Native American: "Respect all Life."

Jainism: "We should regard all creatures as we regard our ownself."

Satanism: "Strive to act with compassion & empathy toward all creatures."

Confucianism: "Don't do to others what you don't want them to do to you."

Buddhism: "One should seek for others the happiness one desires for oneself."

Ancient Kemetic Egyptians, Jews, Christians, Muslims, & Incas all basically agree that: "You shall love your neighbor as yourself."

Found In all world **Religions**

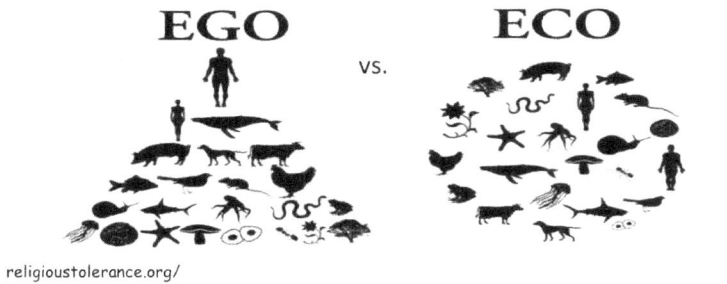

EGO vs. ECO

religioustolerance.org/

Civilization: Perceptions and Reality

Three "Civilized" Perceptions.
Britannica, land of hope and glory.
Belgium, for freedom and law.
United States of America, where the flag stands for freedom.

Reality.
"You can best serve civilization
By being against
What usually passes for it."
So says Wendell Berry ...
And we should all agree!

If could be that many have tried
To serve civilization?
Examples you ask? ...
Dickens' twist on London's exploitations of poor youth,
A very long list.
The Cherokees' trail of tears. Lasted for years. Yet exists my dears.
Helen Hunt Jackson's Ramona underlines the plights, the never-ending fights for indigenous rights.
Conrad's darkness. ... Hearken!
Sinclair's jungle. In a different fashion, but continuance of broad cruelty ... it exists and strangles.
(Old friend and classmate Adolph Chapa got into these bloody, brutal, back-breaking slaughtering tangles
At the ripe old age of fourteen in an Illinois town. ... He was just barely single?)

Ida Wells' battles against lynchings

And fight for the advancement of colored people.
Carson's spring of silencing biocide and the birds that died.
Diamond's guns, germs, and steel.
Man's inhumanity to humanity,
Genocide and slavery were real
And are still.
Catton's overshoot. We have overshot!!!
Hersey's Hiroshima

Oyez! Oyez!
There are many, many more civilization exposés.

Civilization implies a better state. ...
In terms of health of humans and dynamic homeostatic
     symbioses ...
And energetics and hours worked to TRULY LIVE! ...
Sometime, somewhere in the Paleolithic Period
Was the most civilized
While humans have been around.

Currently some one half billion of us
Have 600-plus square feet of air conditioned
Living and automobile space per capita
And transform about 200,000-plus kilocalories per capita
     daily
On the backs of at least three billion humans
While also breaking the back of dynamic homeostatic sym-
bioses
Or Nature.

Perhaps civil was pre or will be post humans and humanity.

Naw.
Can't say that!

I do love self and those of my species too much
To say that.

That's why I propose that we strive to live the seven S's.
And reduce human and domesticated species population
      growth.
And human consumption by Haves.
And increase consumption by have-nots, including other
      species.

Let's live *sabiamente*,
Simply, smally, slowly,
Steadfastly, sharingly,
Sustainably.
As positively ethical applied community ecologists ...
PEACEmakers.

"To live,
We must daily break the body and shed the blood of
      Creation.
When we do this knowingly, lovingly, skillfully, reverently,
It is a sacrament.
When we do it ignorantly, greedily, clumsily, destructively,
It is a desecration.
In such desecration
We condemn ourselves
To spiritual and moral loneliness,
And others to want."
So says Wendell Berry.
And we should all agree!

....

I'm voting, campaigning for Biden and Harris
And perceive I'm doing some good.

But with almost eight billion humans going on ten-plus
    billion
And with too many of these transforming more than
    200,000-plus kilocalories
Of "renewable" and nonrenewable energy each day,
The reality is that civilization will never be civil
Not even with a Bernie and an Elizabeth in the driver's seat.

El mejor nadador es del agua. ¡Sí podemos!

"Lavandera mala no encuentra jamás buena piedra.
A bad washer doesn't ever find a good rock."[1]
(These bad washers blame the system!)

El sistema smells bad!
The system tastes *muy mal.*
It feels bad.
The system appears to be...
So sad!!

"Those Illegal Immigrants
who can no longer
be legally held
(Congress must fix the laws and loopholes)
will be,
subject to Homeland Security,
given to Sanctuary
Cities and States!"[2]

It certainly sounds bad!
Sounds sad!!!
It is sad.
Always has been "bad."

Still ... beauty is in the eye of the beholder.

Stop!
Listen!
Speak!
Hope!
Do!

---

1. *https://en.wikiquote.org/wiki/Spanish_proverbs*
2. Trump Tweet 4/16/2019

¡Sí se puede!
¡Sí podemos!

Swim with.
Swim against.
Swim with and against...
And with.

Antes que te cases,
Mira lo que haces.
Pero...
No confundas hierba con maleza.

Think! Plan!
Y...
¡Deja de rascarte la barriga!
Quit scratching your damn belly!
Plunge in and get wet.
Become a part of it all
As CHANGE!

Swim hard. Fight!
¡Para lo mejor!
¡Para todos!
Los buenos.
Y los malos.
For the better!
For all!

Because you... WE!!!
WE ARE a part of it all.

Respect[1]

"Entre menos burros, más olotes." Dicho mexicano.

"Me? We!!" Muhammad Ali

"The free bird thinks of another breeze ... and he names the sky his own. But
a caged bird stands on the grave of dreams ... his wings are clipped and his
feet are tied so he opens his throat to sing."

Abbreviated from Maya Angelou's Caged Bird

....

Respect within relationships among individuals, demes, populations,
and communities, locally and across the globe, is dependent upon profound and comprehensive comprehension. During my formative years
and afterward, I played, studied, and worked with young and old of various shades of color and various roots and social behaviors in Devine,
College Station, Florida, Georgia, Mexico, Poland, Brazil, and other locales. (During the "Freedom Summer" of 1964 after a feed mill accident,
my life was saved by Afro-Mexican American coworkers, Lacy and Gene
Haywood, probably spurring my quest for respect and justice.) But, I
respectfully acknowledge and confess that despite my good fortune to
have relative power, money, education, resources, and time with which
to work toward improved states of respect, I can be very lazy, and I need
to work much harder at learning how to respect various peoples and assist in the realization of justice for all. To wholly respect others, it takes

---

1. Originally written for a Seguin Gazette series in 2014 initiated by Texas Lutheran
University's Center for Servant Leadership Director, Tim Barr.

a life of learning the nuances of their ethnicities, behaviors, languages, culture, sexual orientations, and history, among others. I also recognize that I need to spend more time communicating with the poorest of this and other regions and countries including ancient (indigenous) and recent immigrants and to afford them with the power, resources, time, and communication skills (in their native language, English, mathematics, through computers and algorithms) to understand me and my culture and history.

As touched upon above, a goal of realizing respect is as simple as the theme of the 2014 prestigious TLU Krost Symposium, environmental justice, i.e., ecological justice, or just justice! It involves simple affirmations that we will do unto, e.g., Haitians and other peoples as we would have Haitians and others do unto to us if we were in their extremely challenging situation. It is simply the achieving significant reductions of consumption by the 1% of humans who consume 75 times per capita what the poorest 20% consume. It would involve drastic reduction of this consumption also because it is decimating populations of other species. Moreover, this simple goal would include the transfer of a large portion of that power of consumption of the 1% over to the 20% who are hungry and undernourished and poorly clothed and sheltered as well as to other species. In addition, it would involve short and long-term management of population growth of humans and domesticated animals. It would necessarily be a process of working toward equality and equity through "applied ecology across curricula and campuses," curricula and campuses of churches, businesses, government entities, not-for-profit non-governmental organizations, as well as school systems, public, private and home schools. It would be the beginning of what I call a generation of positively ethical applied community ecologists/PEACEmakers who have sustainable livelihoods. (Herein, community always includes all species in an area!)

On the other hand, the reality of even the beginnings of a realization of a dream of ubiquitous and universal respect truly is perhaps not so simple but is complex and muddled. It involves grassroot work in the trenches in local communities all over the world as well as heavy-duty politicking at the local level and in regional, state, national, and even international scenes. Certainly, the great works of Sam Flores, A.J. Malone, Dolores Huerta, Willie Velásquez, Elie Wiesel, Betty Friedan, Harvey Milk, W.E.B. DuBois, Rev. Martin Luther King, Jr., Nelson Mandela, Cesar Chavez, Mahatma Gandhi, and others who have fought so hard for respect and justice, but only partially achieved it, have attested to that.

A journey toward truly holistic respect is about positive relationships with other humans and with Nature, and it is about real action. It is not "Yes, sir!" and "No, sir!" to the status quo, including our current world systems' status quo of rampant transformation of energy and subsequent loss of existing topsoil, quality of water systems, and biodiversity. Respect is not bowing down to the power structures who worship big and fast at the expense of others including other species. It is most certainly not a neglect of going to the voting booth and a disdain for getting involved in the political process. An attitude of respect does not involve the acceptance of our current socio-political/economic systems, which are so exploitive of relatively stable natural economies. Respect is looking hard for a different route toward individual quality life which is quality life for all the community. It starts by speaking truth to power and to apathy and to ignorance of ecological processes and principles. (Global and local power might include the moneyed, transnational corporations, politicians and bureaucrats, the majority in a democracy, the military-industrial complex, or religious institutions.)

Down these lines, I have to say that too many of us, including yours truly, are often quite satisfied with our sheltered "life" in comfortable artificial bubbles. However, surely we should begin to more fully live and launch a critical mass of relatively wealthy, powerful and "good" Christians /Muslims /Jews /Hindus /Buddhists /humanists /others who would travel to troublesome sites in great numbers and use strategies and tactics of civil disobedience and non-violence to stop wars in Syria, Central Africa, Ukraine, and other areas of the world. Similar tactics and strategies can be utilized to curb continued production of weaponry, extreme poverty in Zimbabwe and Haiti and other countries, ecological disruption all over the globe, and perhaps one day take us a bit closer to open borders and cosmopolitan and relative peace.

Respect must start with self, but self must also include local and global community. Respect is an effort toward pristine and natural and is not polluting and not so artificial. It recognizes anthropogenic detrimental changes in Nature/the Land. We are but one of many species working at survival, but we are the dominant one on this ecosphere and we need to recognize that an excess of this dominance can lead to extinction of quality life for more and more. Respect is having Faith in the power of good and community, but respect also gives Science equal weight and works to conserve and share with others the resources of mineral and hydrological cycles, photosynthesis and biodiversity and the daily solar energy that arrives on this ecosphere. (On the other hand, it is not imprudently and totally getting on the bandwagon of STEM and many of the values being pushed in these programs. In some ways this currently fashionable "education" process of STEM has little respect for positively ethical applied community ecology or faith in Nature/the Commons/the Land, and is far too focused on faster and bigger, pseudo-growth, inappropriate technology, upward mobility, and money along with power over resources.)

As far as the Latin derivation of the word respect is concerned, in looking back I appreciate any good/god I may have that has been instilled in me from the villages in which I have lived including Seguin, Texas, U.S.A. I am grateful for the freedom of continuing development these villages provided or are now providing toward eliciting freedom songs from "caged birds" and enabling "free birds" to realize "We!"

Finally, I want to get back to the point made in the third (3rd) paragraph herein that respect is a complicated and confused process. It is an understatement to say that humans and human relationships are not perfect and that change toward real conservation, resilience, and sustainability, i.e., social justice, humaneness, and ecological sanity, is tough and incredibly challenging. But across the ecosphere much does exist in the way of healthy relationships, dialogue, discussion, listening, diplomacy, and consensus-building. And we do need to shed plenty of light on this good which is taking place in community. Indeed, there is much happening in many communities that might get us on a road toward increased respect and good-/god-liness, i.e.:

--voter registration drives

--some research into various communities' sustainability status (and committees on sustainability)

--dialogue about better and more holistic educational systems including continuing ecological education

--initiatives concentrating on more art, historical knowledge, learning of other languages, including mathematics and computer languages

--programs to increase the physical and spiritual health of individuals/family/community/…"

--scholarships, internships, and leadership programs focusing on the less fortunate

--naturalists working to protect and enhance green space

--programs to achieve more nutritious, energetically-sound and user-friendly food systems

--construction of libraries in concert with goals of the long-term conserving of energy and resources

--some efforts at reducing, reusing and then recycling

--efforts at mass transport and at realizing bicycle lanes and increased bicycling for transport rather than just for exercise or recreation

--efforts toward quality life in Honduras, Mexico, Haiti, Africa and other parts of the world

--realization that landfills and waste can be very problematic

--perhaps some realization that (conspicuous) consumption and the desperate grab for and transformation of energy can be a serious problem

--the respect campaigns such as the one which initiated this article and which robustly attempted to stimulate communication within the community within and around Seguin, Texas, during much of the year of 2014 and onward!!

"To live, we must daily break the body and shed the blood of Creation. When we do this knowingly, lovingly, skillfully, reverently, it is a sacrament. When we do it ignorantly, greedily, clumsily, destructively, it is a desecration. In such desecration we condemn ourselves to spiritual and moral loneliness, and others to want."

Wendell Berry

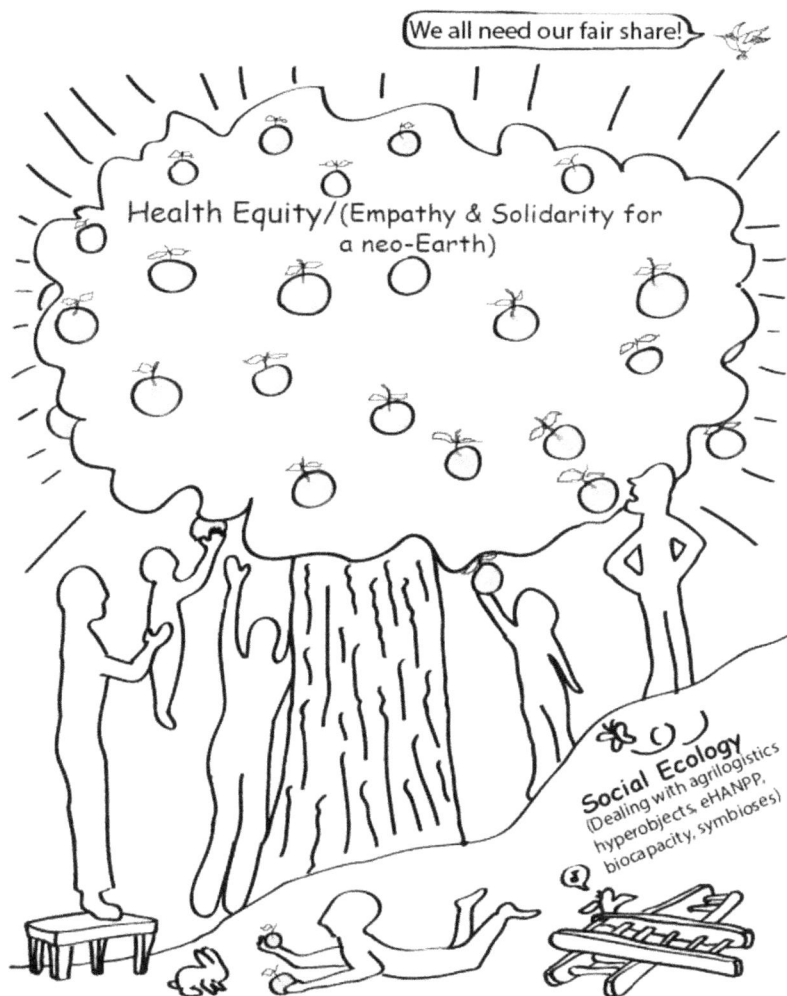

Why? Health care needs to be about

(1) sustainability (social justice, humaneness, ecological sanity) of all life and life systems, and

(2) sustainable livelihoods (equality/equity in world populations and health care practitioners/providers who are more about profound, holistic ecological health than about personal wealth and conspicuous consumption and self-accumulation of power [money, stuff, land], and who wish to live in concert with nature [symbioses].)

> If I and my habitat aren't healthy, then neither are you and yours!

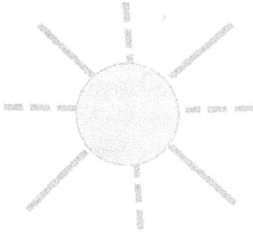

Sustainable Livelihoods

"All education is environmental education."
David Orr, Ecologist, Oberlin College

Background. The real crisis in the world is not in the financial economy and its current state, but rather, the crises in Nature's economy (severe human poverty and malnutrition [and other physical and mental/spiritual stresses on humans], watershed disruption, top soil loss, dead zones, desertification, loss of diversity and resilience, serious pollution, global climate change, etc.). Moreover, we are not providing our children or most of the "adults" around them with the educational foundation for developing critical thinking and ethical decision-making skills, particularly regarding the serious long-term ecological challenges.

(Of course, in the U.S.A. most of the problems with Nature's economy are out of sight and mind because we have so much power, and we suck tremendous resources from all over the world to a relatively small population here in North America. The process masks and hides the serious problems our kids and grandkids will have to confront with insurmountable difficulties. Moreover, most of us live in such a virtual and unreal reality that whole lives of relative ignorance and procrastination are prevalent versus what could be fulfilling and spiritually rich and active lives wisely dealing with real problems.)

The bottom line is, for the most part sustainable livelihoods do not exist in our very artificial conventional economic systems. Our current and past economies, and most of our livelihoods that come from these socio-political/economic systems, are destroying soils, water, the air, and the climate, all of which sustain life. These unsustainable livelihoods are doing away with the organisms and their ecological communities with which we as humans must associate for quality life. They are destroying our humaneness, our humanity!

Some of us believe we should earnestly begin to attempt to change this unpleasant situation we humans are creating because of our development and continued propping up/bandaiding of non-conserving, unsustainable, and non-resilient ecological communities. In particular, we are certain that this major shift in behavior and action must include a comprehensive and intensive long-range plan, which would involve optimally small (less that a 500 student population) neighborhood and rural schools—with separate elementary, middle school and high school campuses placed side by side, but in concert with the Land and Nature and with as many classes as possible being held outdoors.

"To live, we must daily break the body and shed the blood of creation... when we do this knowingly, lovingly, skillfully, reverently, it is a sacrament; when we do it ignorantly, greedily, clumsily, destructively, it is a desecration."

Wendell Berry, Essayist, Poet, Farmer

Sustainable Livelihoods. We are positive that such an ecologically-sound school system mentioned in the previous section can help to realize sustainable livelihoods for local communities and the world, livelihoods which involve some of the following:

- Educated holistic and ethical decision-makers

- Folk who dedicate their lives to focusing on the poor with eduation, knowledge, franchisement, empowerment, power, and resources

- Organic farmers who are "truly organic" in a holistic sense

- Urban farmers and rural farmer-ranchers who produce grass-fed and browse-fed meat animals on a small and large scale

- Holistic low-input community gardeners

- Health care professionals who holistically and comprehensively practice preventative care on a local level and curative care when needed and who develop health care systems that particularly focus on the poor

- Lawyers who mostly help the poor (including other species that are poor as far as power is concerned)

•   Bankers supporting microloan/microenterprise systems which are conserving and sustainable

•   Blue collar workers who make enough for a good quality life

•   White collar workers who make enough for a good quality life but no more

•   Architects who design conserving and sustainably built-systems

•   Builders of small, ecological-friendly homes

•   Constructors and maintainers of transport systems primarily involving bicycles, trains, buses, and modern clipper ships

•   Seekers of low input/throughput/output systems involving ethical use of what is truly "renewable energy"

•   Effective and efficient communicators who work in inexpensive, low input systems

•   Systems analyzers and researchers who can effectively communicate the state of the state/the world in terms of material flow and energy flux inputs, throughputs, and outputs; also, teams producing life cycle assessments for products/systems

•   Scientists who truly seek knowledge vs. technicians and technologists who attempt to bring the Land/Nature "to its knees" in service to humans

•   Guardians of diverse, native, living communities of organisms (including in bays and estuaries); ample amounts of good clean water and air; rich, deep, living top soils; and ethical use of energy

•   Ethical naturalists

•   Readers who seek socio-political/economic (ecological) knowledge about how to live well in a place

•   Human cultures that respect other human cultures, traditions, and rituals

•   A human culture that respects the Nature, the Land

- Ecological historians

- Local, homegrown entertainers (musicians, singers, dancers, co-medians) who are relatively "low input"/"low maintenance"

- Everyone actively participating in local, low maintenance sports and entertainment

- Politicians and bureaucrats/policy-makers at all levels who work intelligently and prudently to facilitate change toward "conservation and development of sustainable community"

- Teachers of reading, writing, and arithmetic who are striving to meet our local and global challenges within a holistic, participatory/hands-on, site-based curriculum of applied ecology

- True Peacemakers

- Folk in all disciplines and roles in life who are Positively Ethical Applied Community Ecologists and who live light on the Land

"My own preference is for an environmentalism that talks about ethics and aesthetics rather than about resources and economics, that places priority on the survival of the living world of plants and animals regardless of their productive value, that cherishes what nature's priceless beauty can add to our deeper-than-economic well-being."

Don Worster, Environmental Historian, University of Kansas

<u>Why?</u> Do We Need Science? (The systematic quest for knowledge/
the search for Truth.)

> Science is reductionistic, holistic, and with a systems-approach. It is dynamic and utilizes a consensus process among good, reputable scientists.

Traditions and Myths Are Important for
Stability and Peace, and a Holistic Ethic of Reciprocity.

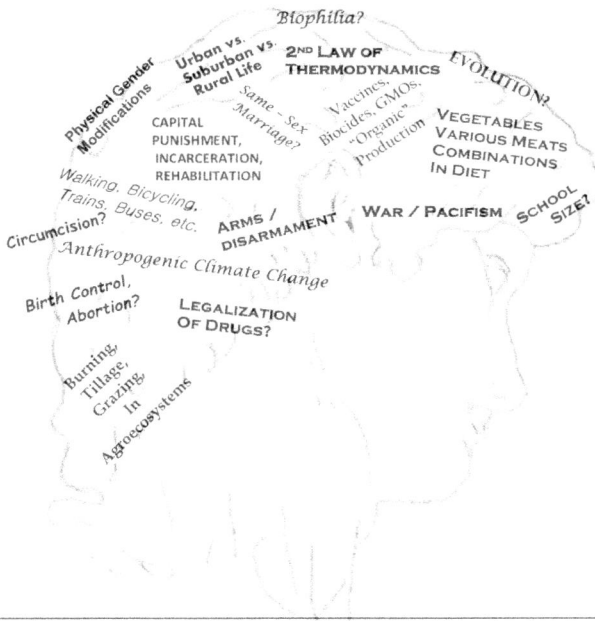

Biophilia?

Physical Gender Modifications

Urban vs. Suburban vs. Rural Life

2ND LAW OF THERMODYNAMICS

EVOLUTION?

Same - Sex Marriage?

Vaccines, Biocides, GMOs, "Organic" Production

VEGETABLES VARIOUS MEATS COMBINATIONS IN DIET

CAPITAL PUNISHMENT, INCARCERATION, REHABILITATION

Walking. Bicycling. Trains. Buses. etc.

Circumcision?

ARMS / DISARMAMENT

WAR / PACIFISM

SCHOOL SIZE?

Anthropogenic Climate Change

Birth Control, Abortion?

LEGALIZATION OF DRUGS?

Burning, Tillage, Grazing In Agroecosystems

However, they must be based on "current" scientific (ecological)
knowledge, the Precautionary Principle, the 2nd Law of
Thermodynamics, respect for Nature,
and a holistic Ethic of Reciprocity.

Philosophy and Ecological Ideas

(Thoughts generated during and from a weekend of philosophy and ecology at Rice University's Moody Center, January 26, 2019 organized by the Institut français, Paris, attendance of a service at an evangelical/charismatic church, breakfast and lunch with dear family, and books and articles around me)

Philosophy—The study of the fundamental nature of knowledge, reality, and existence, especially when considered as an academic discipline. *https://en.oxforddictionaries.com/definition/philosophy*

....

o **I am in essence a (perhaps wannabe) campesino/hunter-gatherer**[1] (Maybe we all are.) I have discomfort with too much order, luxury, too much artificiality, built-environment, industrialization, and electrification and the resultant chaos in larger wholes which results from ordered states in their sub-wholes (i.e., the business of the Second Law of Thermodynamics/energy quality, energy flux, entropy).

o **The Meaning of Human Existence, E.O. Wilson.** "The unfolding of history is obedient only to the general laws of the Universe. Each event is random yet alters the probability of later events."

"We are not predestined to reach any goal, nor are we answerable to any power but our own. Only wisdom based on self-understanding, not piety, will save us. There will be no redemption or second chance vouchsafed to us from above. We have only this one planet to inhabit and this one meaning to unfold."

---

1. Sometime in the Paleolithic Period may have been the best of times for *Homo sapiens*, and agriculture/agrilogistics may have been the worst mistake of humankind. Diamond, Jared 1999 The worst mistake in the history of the human race. Discover, May 1

"[We] have saturated a large part of the Earth [Eaarth], and altered to varying degree the remainder. We have become the mind of the planet and perhaps our entire corner of the galaxy as well. We can do with Earth what we please. We chatter constantly about destroying it—by nuclear war, climate change, and apocalyptic Second Coming foretold by Holy Scripture.

Human beings are not wicked by nature. We have enough intelligence, goodwill, generosity, and enterprise to turn Earth into a paradise both for ourselves and for the biosphere that gave us birth. We can plausibly accomplish that goal, at least be well on the way, by the end of the present century. The problem holding everything up thus far is that *Homo sapiens* is an innately dysfunctional species. We are hampered by the Paleolithic Curse: genetic adaptations that worked very well for millions of years of hunter-gatherer existence but are increasingly a hindrance in a globally urban and technoscientific society."

"We are addicted to tribal conflict, which is harmless and entertaining if sublimated into team sports, but deadly when expressed as real-world ethnic, religious, and ideological struggles. There are other hereditary biases. Too paralyzed with self-absorption to protect the rest of life, we continue to tear down the natural environment, our species irreplaceable and most precious heritage. And it is still taboo to bring up population policies aiming for an optimum people density, geographic distribution, and age distribution. The idea sounds 'fascist,' and in any case can be deferred for another generation or two—we hope."

"Our leaders, religious, political, and business, mostly accept supernatural explanations of the human existence. Even if privately skeptical, they have little interest in opposing religious leaders and unnecessarily stirring up the populace, from whom they draw power and privilege. Scientists who might contribute to a more realistic worldview are especially disappointing. Largely yeomen, they are intellectual dwarves content to stay within the narrow specialties for which they were trained and are paid."

o **Humankind: Solidarity with Nonhuman People, Timothy Morton.** "Ecological awareness is knowing that here are a bewildering variety of scales, temporal and spatial, and that the human ones are only a narrow region of a much larger and necessarily inconsistent and varied scalar possibility space, and that the human scale is not the top scale."

"Thinking about action this way is superior to actor-networks or the higher volume version, mechanical pushing around, which is the scientistic version of Neoplatonic Christianity, the thing that even Descartes (who says he isn't) is retweeting, and the thing that Kant (who says he isn't making the same mistake as Descartes) is also retweeting. This bug has affected many thought domains. Industrial capitalism is theorized by Marx as an emergent property of industrial machines—when you have enough of them, pop! But this means capitalism is like God, always greater than the sum of its parts."

"Philosophy requires a new theory of action, a queer one that is neither active nor passive nor a compromised amalgam of both, to help us slip out from underneath physically massive beings such as global warming or neoliberalism, to find some wiggle room down there so we can wriggle or rock our way out of the hyperobjects."

"Love is not straight, because reality is not straight. Everywhere, there are curves and bends, Things veer. Per-ver-sion. En-vir-onment. These come from the verb 'to veer.' To veer, to swerve toward: am I choosing to do so or am I being pulled? Free will is overrated. I do not make decisions outside the Universe and then plunge in, like an Olympic diver. I am already in."

"It's not just that you can have solidarity with nonhumans. It's that solidarity implies nonhumans. Solidarity requires nonhumans.

Solidarity just is solidarity with nonhumans."

o **"Self-Help, Ancient Greek Style: Aristole offers a way forward to a well-balanced and happy life.**" By John Kaag. NYT Book Review, Jan 27, 2019 "Living a virtuous life, for Aristotle, comes down to: 'Nothing in excess.'"

o **Institut français' and Moody Center for the Arts' Night of Philosophy and (Ecological) Ideas, Jan 26**[2] Laurie Anderson and Timothy Morton. It's Not the End of the World. That Was a While Ago—Laurie and Tim discussed ecology or key questions for humanity. They stated that if you think technology is going to solve problems then you do not understand technology. Even more important is that you do not understand the problems!

Nevertheless, don't panic.

---

2. See next note, page 173.

I'm in a Hurry (and I Don't Know Why)

Roger Murrah and Randy VanWarmer

"I'm in a hurry to get things done
Oh I rush and rush until life's no fun
All I really gotta do is live and die
But I'm in a hurry and don't know why."

And, do not ignore the sad but have hope. You cannot live without hope. Something else will rise. All is love/a desire to be free. Even suicide.
What happens to karma if all becomes gone? Well, there are/will be other universes:

The Second Coming

William Butler Yeats

"Turning and turning in the widening gyre
The falcon cannot hear the falconer;
Things fall apart; the centre cannot hold;
Mere anarchy is loosed upon the world,
The blood-dimmed tide is loosed, and everywhere
The ceremony of innocence is drowned;
The best lack all conviction, while the worst
Are full of passionate intensity.

Surely some revelation is at hand;
Surely the Second Coming is at hand.
The Second Coming! Hardly are those words out
When a vast image out of Spiritus Mundi

Troubles my sight: somewhere in sands of the desert
A shape with lion body and the head of a man,
A gaze blank and pitiless as the sun,
Is moving its slow thighs, while all about it
Reel shadows of the indignant desert birds.
The darkness drops again; but now I know
That twenty centuries of stony sleep
Were vexed to nightmare by a rocking cradle,
And what rough beast, its hour come round at last,
Slouches towards Bethlehem to be born?"

Talk to sad, panic, hopelessness.

The Sounds of Silence

Paul Simon

"Hello darkness, my old friend
I've come to talk with you again
Because a vision softly creeping
Left its seeds while I was sleeping
And the vision that was planted in my brain
Still remains
Within the sound of silence
In restless dreams I walked alone
Narrow streets of cobblestone
'Neath the halo of a streetlamp
I turned my collar to the cold and damp
When my eyes were stabbed by the flash of a neon light
That split the night
And touched the sound of silence
And in the naked light I saw
Ten thousand people, maybe more
People talking without speaking
People hearing without listening

People writing songs that voices never share
No one dare
Disturb the sound of silence
'Fools' said I, 'You do not know
Silence like a cancer grow
Hear my words that I might teach you
Take my arms that I might reach you'
But my words like silent raindrops fell
And echoed in the wells of silence
And the people bowed and prayed
To the neon god they made
And the sign flashed out its warning
In the words that it was forming
And the sign said 'The words of the prophets
Are written on subway walls
And tenement halls
And whispered in the sounds of silence'"

I no longer believe in history:

Strange Angels

Laurie Anderson

"They say that heaven is like TV
A perfect little world that doesn't really need you
And everything there is made of light
And the days keep going by
Here they come
Here they come
Here they come.
Well it was one of those days larger than life
When your friends came to dinner and they stayed the night
And then they cleaned out the refrigerator
They ate everything in sight

And then they stayed up in the living room
And they cried all night
Strange angels singing just for me
Old stories they're haunting me
This is nothing like I thought it would be.
Well I was out in my four door with the top down.
And I looked up and there they were,
Millions of tiny teardrops just sort of hanging there
And I didn't know whether to laugh or cry
And I said to myself, What next big sky?
Strange angels singing just for me
Their spare change falls on top of me
Rain falling
Falling all over me
All over me
Strange angels singing just for me
Old Stories they're haunting me
Big changes are coming
Here they come
Here they come."

"Living in the Future's Past"—This is a well-made 2018 film with Jeff Bridges about the why of PEACE/Positively Ethical Applied Community Ecology. *https://www.livinginthefuturespastfilm.com/* It deals with Reality 101, or "everything you need to know about reality so you don't spend the rest of your life in total stupidity" by Robert A. Jacques.

A fantastic study guide is at *https://www.videoproject.com/assets/images/ PDF/Living_in_the_Futures_Past.pdf*

"Not OK"—This little movie is about a small glacier at the end of the world and it is also about the why of PEACE. Perhaps it is not as profound and comprehensive as other films presented, but it is a thought-provoking 2018-film! *http://worldfilmpresentation.com/film/not-ok-little-movie-about-small-glacier-end-world https://www.notokmovie.com/*

"Tomorrow"—This 2015 film is about the what and how of PEACE. *https://www.tomorrow-documentary.com https://www.nytimes.com/2017/04/19/ movies/tomorrow-review.html*

Titles of the breakout sessions--Experiencing Nature: Arousal, Interest, and Aesthetics; To Cohabit with Wild Nature, Is It Impossible; The Wild and Wicked: Why You Don't Have to Love Nature to Be Green; Awakening Concern: the Art and Science of Repairing, Mending and Making Things Right; Is There a Moral Obligation to Go to Mars?; Environmental Threats and What Justice Demands; The Philosophy Behind the Trump Administration's Energy Agenda.

Creators and names of exhibits—Natsha Bowden: Sideways to the Sun; Jae Rhim Lee: Infinity Burial Suit Project; Michel Blazy: We Were the Robots; Momoko Seto: Planet ∞ ; Justin Brice Guarigilla: We Are the Asteroid III.

o **The Epoch Times.** In my hotel in Katy, Texas I found a pile of The Epoch Times, a right of center newspaper which in the January 10-16, 2019 edition was spreading fear of immigrants, Sharia law, leftists, Barack Obama, Sr., Democrats, & regulations; touting the great feats of white Afrikaans; and congratulating Trump. *https://en.wikipedia.org/wiki/The_Epoch_Time*

o **Bayou City Fellowship (Evangelical/Chrismatic) Sunday Service with a lovely & loving niece.**[3] I've experienced these types of services throughout my life in Pentecostal & southern black Baptist churches, at the botanical gardens in Rio, in rural Nicaragua, and even to some extent in the Roman Catholic High Masses in Latin when I was a kid:

Tantum Ergo

Thomas Aquinas

"Tantum ergo Sacramentum
Veneremur cernui:
Et antiquum documentum
Novo cedat ritui:

---

3. Hypocrisy ("A feigning to be what one is not or to believe what one does not : behavior that contradicts what one claims to believe or feel." Merriam-Webster) was discussed at both the Moody Center- and Bayou City-Fellowship gatherings. I believe that there is plenty of hypocrisy to go around amongst humanity—so-called ecologists and so-called christians, etc.

Praestet fides supplementum
Sensuum defectui.
Genitori, Genitoque
Laus et jubilatio,
Salus, honor, virtus quoque
Sit et benedictio:
Procedenti ab utroque
Compar sit laudatio. Amen."

"Down in adoration falling,
Lo! the sacred Host we hail,
Lo! o'er ancient forms departing
Newer rites of grace prevail;
Faith for all defects supplying,
Where the feeble senses fail.
To the everlasting Father,
And the Son Who reigns on high
With the Holy Spirit proceeding
Forth from each eternally,
Be salvation, honor blessing,
Might and endless majesty. Amen."

With so many desperately searching for the meaning of human existence, these evangelical/charismatic services seem to be very popular around the world. (By the way, my journey and searching has brought me to perhaps being an agnostic-ignostic continuing-learner, and my god may be "dynamic-homeostatic-symbioses".)

All is connected. There is a connection in what was taking place at the Moody Center from 7:00 pm January 26, 2019 until 1:00 am January 27, 2019, [or the Night of Philosophy & (Ecological} Ideas] and what was occurring at Bayou City Fellowship from 9:00 until 10:30 am January 27, 2019.

Asks the preacher, … "What is Christian/christian faith?" The young fellow didn't answer his question. I hope his intention in asking this question was to provoke the seeking of good quality life for

180

all, including other species, for as long as possible. I trust that that being christian is doing right for all in the ecosphere. If this is what it is, then christians need to understand the science of limits to the natural resource base, of exceeding the carrying capacity of the Earth, of disparity, and of a real need for empathy and an ethic of reciprocity (a profound, comprehensive, and holistic golden rule), and of abiding by the precautionary principle. And, I hope he recognizes that life/"nature"/symbioses is a zero sum game. Finally, I trust that he recognizes a desperate need for ecological literacy.

The pastor mentioned that the congregation would learn over time how to interpret Scripture, and he mentioned difficulties with Leviticus but did not go into this. Perhaps his explanation might be something like this, "Some people say, 'You Christians pick and choose which verses of the Bible you want to obey. The book of Leviticus prohibits homosexuality and what you can eat and prescribes animal sacrifices. Why do you disregard some rules but adhere to others?' There is a simple answer. The only rules of the Old Testament that apply to us today are the rules that are repeated in the New Testament. We don't live under the old law. We live under the new law of God. The New Testament says nothing about dietary restrictions or animal sacrifices, but it does repeat the commands about adultery, premarital sex, and homosexuality." Or he might have fabricated other desired-excuses and -dogma in some other way to dictate unnatural social mores not based on science.

As the pastor read or referred to scripture and mentioned the Book of Genesis, I thought about anthropologist Hugh Brody's book, The Other Side of Eden. In his 2001 review in The Guardian of The Other Side of Eden, Ros Coward writes:

"Normally, hunter-gatherers are seen as nomads and farmers as settlers. Brody thinks the reverse is true. Farming culture is accompanied by 'a longing to be settled, a defensive holding of ground and a continuing endemic nomadism' caused by the continuous growth of population among such communities. Genesis, says Brody is the ultimate agriculturalist myth, embodying their continuing quest to reshape nature as a lost Eden. Hunter-gatherers, by contrast, do not seek to reshape and dominate their landscape. Their conviction is that their land is 'already Eden and exile must be avoided'. "

"As well as being an argument for the political rights of hunter-gatherer societies, The Other Side of Eden is also a passionate argument in support of recognising and nurturing the hunter-gatherer world-view. At a time when nature is so under threat from humanity, there are invaluable environmental lessons to be learnt from cultures which seek to survive from the land but also leave it as they find it." *https://www.theguardian.com/books/2001/jan/28/society*

How? Traditions and faith vs. conventional science. Part 1 of 2
(Science-Based) Traditions/Faith

Taboos on eating insects, catfish, meat, beef, pork, humans, ... Much of this can make sense.

"If it ain't broke, don't fix it!" Cultural mores, celebratory activities, traditions, faith, religion can be the glue for a sustainable ecological community involving humans. Besides, we are learning little through science, and we need to always remember to abide by the Precautionary Principle.

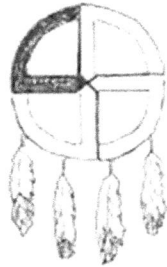

How? Traditions and faith vs. conventional science.
Scientific-Based Change/Open-mindedness

Human organizational structures/cultures/mores/religions
are never perfect, and there are things which need fixing and
changing, i.e., through caution and tentativeness, and based on
knowledge.

THE NATURAL PHENOMENA
(Sustainability should be realized through science as well as tradition.)

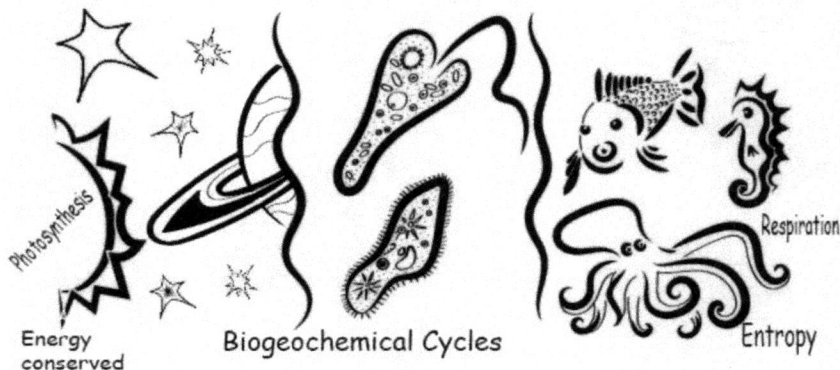

Photosynthesis

Energy
conserved

Biogeochemical Cycles

Respiration

Entropy

Why? In the past and present, disparity among humans and among life-forms was/is a major challenge. (The challenge is particularly great in today's world!)

> Chirp, chirp. ... I realize it is sort of "natural" for you Haves to be smug. But do you really want to do without us and lose a healthy diversity of biota?

Ca. 2020:

1% have power over most of the resources of the Earth!!!

0.5-1 billion have far too much!!

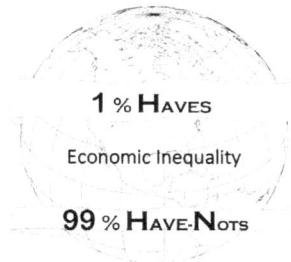

1 % Haves

Economic Inequality

99 % Have-Nots

3 billion are in relative poverty ...

including many farm workers and other laborers in China, Central America, the Philippines, Africa, and India, etc., and many bargain-store type employees.

1 billion in poverty currently (2020) have it worse, or nearly as bad, as at any time in the history of humankind.

Earth ⟶ Eaarth = Earth – Nature

Habitat for many of the other 9 million species is being destroyed and we have an unprecedented extinction rate.

Why? We Need a Totally New Socio-Political/Economic (Ecological) System

1. In today's dynamic high-input/-throughput world we are *de facto* dependent on/addicted to fossil energy. Our rampant transformation of energy causes chaos & stress, pollution and destruction of living systems, and global climate change.
2. If we transform "renewables" rapidly at too high a rate, they will "run out" and thus are not "renewable." (There is only so much biomass, daily available solar energy, wind, water flow, accessible geothermal.) More important is that it takes considerable fossil energy to capture and transform diffuse/low-quality "renewable energy."
3. Nuclear and fossil energy have too many pollution/negative externality red flags.
4. The solution is to slow our economic engine down (to reduce the collective ecological footprint, population growth, consumption) and rely on the ultimate "renewable," i.e., daily solar energy.

> I'm natural & perpetual. The "Energizer Bunny Rabbit" has to have constant injections of unsustainable fossil energy

A Simple View of the Energetics of the Human Economy

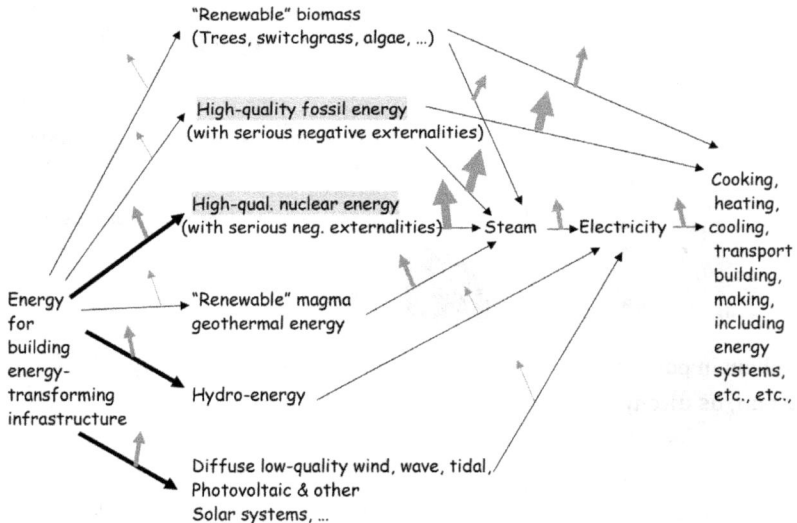

"Renewable" biomass
(Trees, switchgrass, algae, ...)

High-quality fossil energy
(with serious negative externalities)

High-qual. nuclear energy
(with serious neg. externalities) → Steam → Electricity

Energy
for
building
energy-
transforming
infrastructure

"Renewable" magma
geothermal energy

Hydro-energy

Diffuse low-quality wind, wave, tidal,
Photovoltaic & other
Solar systems, ...

Cooking,
heating,
cooling,
transport
building,
making,
including
energy
systems,
etc., etc.,

Above ... quantity of "entropy," "chaos," "heat pollution," "loss of biodiversity" is indicated by the number and size of the arrows.   (↑)

186

Renewable Energy as THE Key Asset of Commonwealth in Community

Renewable energy[1] should be ethical, just transformation of energy in a low-input/-throughput steady-state, human, economic system. Renewable energy sources "capture their energy from existing flows of energy, from on-going natural processes, such as sunshine, wind, flowing water, biological processes (e.g., photosynthesis, etc., into biomass), and geothermal heat flows. The most common definition is that renewable energy is from an energy resource that is replaced rapidly by a natural process such as power generated from the sun or from the wind. Most renewable forms of energy, other than geothermal (from magma) and tidal power, ultimately come from the Sun."

Life is matter organized with inputs of free, available energy. Therefore, energy is the glue of all living systems, systems that may be cells, organisms, ecological communities (including those which are very artificial subsets of Nature), the Land, and Commons-components. However, inappropriate and/or excessive transformation of energy results in life systems becoming unglued and unhealthy.

Basic truths (all profoundly connected) for a resilient and sustainable Earth are as follows:

biophilia and a need for Nature,

an ethic of reciprocity (which, in a more profoundly holistic interpretation, includes all living organisms/Nature), and

limits and carrying capacity (largely because of the 2nd Law of Thermodynamics, which rules that use/transformation of the perpetual amount of energy results in this energy becoming relatively useless, and excessive transformation of it creates more general chaos in larger or other wholes [of Nature]).

---

1. Renewable energy is, in essence, an oxymoron since "Energy cannot be created or destroyed" (the 1st Law of Thermodynamics) and "As energy is transformed it tends toward uselessness" (the 2nd Law of Thermodynamics).

These truths guide us in processes of just transformation of energy.

Transformation of the 2% of free, available solar energy captured by photosynthesis resulted in a dynamic but relatively homeostatic biosphere over 4+ billion years of evolution. This homeostasis continued even into much of the 200,000+ years in which humans became a part of the ecosphere. However, in the last 12,000 years or so, and particularly the last 70, this homeostatic situation is being seriously challenged. Major challenges and perturbations have been from

- the agriculture revolution,

- industrialization and increasing use of fossil energy and products from fossil material/energy such as plastics, nitrogenous molecules/ fertilizers, pesticides, pharmaceuticals, concrete, asphalt and

- the microelectronic and information revolution.

In addition further issues exist with concurrent growth of human populations (projected to reach 10 billion in 2050)[2] and energy consumption or transformation (currently at up to >250,000 kilocalories/ capita daily, or up from 2,000 for "primitive" humans and 70,000 for early industrial humans)[3]

Therefore, in not limiting growth of human consumption and human populations, we are disrupting the "homeostasis" of the various assets of commonwealth. (Discussion of twelve assets of commonwealth as proposed by the NGO, Ogallala Commons is at *https://ogallalacommons. org/about-us/commonwealth/*)

We are in a profoundly serious way negatively affecting the wellness of individuals and populations of lifeforms of Eaarth. Real and sustainable enhancement of each of the assets of commonwealth depends on ethical/just transformation of energy, i.e., renewable energy, and living individual and collective human lives

Sabiamente (Spanish for knowledgeable and wisely, or prudently),

Simply, Smally, Slowly,

Steadfastly,

---

2. Human populations did not reach 1 billion until 1804.

3. *https://www.wou.edu/las/physci/GS361/electricity%20generation/HistoricalPerspectives.htm*

Sharingly,

Sustainably.

This may be somewhat of a bitter pill for those who believe in boundless opportunities, across the Great Plains and elsewhere, for artificialization (growth of consumption and human populations). On the other hand, there are innumerable satisfying and self-fulfilling opportunities for facilitating learning about "positively ethical applied community ecology/PEACE" across campuses of all human entities and developing steady-state human economies involving low-inputs/-throughputs. Implementation of this process toward resilient, sustainable ecological community is desperately needed!

If we consider renewable energy to be ethical transformation of energy in ecological community, we can make some broad generalizations

1. Decisions about use of nuclear energy and dirty fossil energy (especially coal, peat, shale, tar sands, heavy and extreme oils) are basically no-brainers. They are not renewable energy! They have too many red flags of unintended consequences and negative externalities.[4] They should be avoided!

2. "We can't get energy for nothing; it takes energy to get energy." Moreover, there are always unintended consequences and some negative externalities which come with actions taken to transform energy. Therefore, we must always be cautious and tentative when converting energy whether it is high-quality relatively non-renewable fossil energy or nuclear energy, or whether it is relatively diffuse, lower-quality renewable energy.

3. Everything runs on moderate- to high-quality energy that cannot be recycled, so choose energy resources wisely. (We are beginning to get a positive energy return on investment from transformation of what are currently considered to be renewable energies [windticity, photovoltaics], even though they are generally diffuse and relatively low- quality energies. However, the research, development, utilization, and maintenance of these "renewable" energy systems take inputs or embodied high-quality energy such as various forms of fossil energy and fossil materials. Therefore, remember that we may never actually get a positive energy

---

4. A cost that affects a party who did not choose to incur that cost." *https:// en.wikipedia.org/wiki/Externality*

return on invested energy when we go after diffuse sources of energy and there are often negative externalities involved. My analyses of this situation always take me to a need for livelihoods and lifestyles which are "simple, small, and slow".)

4. Heating and cooling using geothermal systems (ground source heat pumps or magma heat sources), hydroelectric units, windtricity, photovoltaic systems can be relatively suitable renewable energy sources as long as there is real net energy in the transformation without too much fossil fuel or nuclear energy invested in them, and without too much ecological (including psychological, socio-political, and economic) destruction.

5. However, the best renewable energy process is photosynthesis in natural ecosystems as well as through appropriate applied agroecology, along with low-input human systems involving appropriate landscaping, insulation, dog-runs, and other passive solar heating and cooling strategies such as clothes lines and solar dryers and cookers (and more walking and bicycling). Natural photosynthetic systems, in which humans are in relative concert with Nature,[5] are relatively resilient, self-sustaining/perpetual.

It is of unethical hubris to believe that humans can do better holistically, profoundly, resiliently, and sustainably than the 4+ BILLION years of evolution of photosynthesizing Nature. We need to wean ourselves from high-input/-throughput systems that are disruptive of healthy ecological communities across the Great Plains and elsewhere. Necessary also is a rapid and smooth transition to protection of and judicious use of net primary productivity[6] in relatively natural systems.

And, this limits us to human systems which are simple, small and slow![7]

Now to somewhat reiterate with a question, what does "simple, small, and slow" (de facto renewable energy systems) mean for the everyday lives of Jane or John Q. Public? Keeping families/homes small is

---

5. This means curbing population growth and numbers of humans and domesticated species and especially curbing per capita consumption.

6. The rate at which an ecosystem accumulates energy or biomass, excluding the energy it uses for the process of respiration. This typically corresponds to the rate of photosynthesis, minus respiration by the photosynthesizers.

7. I propose a goal of a relatively equitable distribution of 70,000 kilocalories/capita daily for a possible stable human population of ca. 10 billion in 2050 and beyond.

a big and pragmatic step. And since food is the largest or one of the largest components of our ecological footprint, raising the right kind of foodstuffs locally in an ecologically-sound way helps to take us down a moral and ethical road toward sustainability. Finally, one of the most appropriate forms of technology is the bicycle. It should be used extensively and regularly, and it and terrestrial-mass transport such as trains must replace cars, semi-trailer trucks, and other automobiles. (Of course, there are many other little livelihood and lifestyle habits which can be considered and adopted on a road toward ethical/just transformation of energy or low-input/-throughput, resilient and sustainable community.)

Why? How? Achieving "Positively Ethical Applied Community Ecology"

(in all communities)   Let's do it!!!!

1. Develop a foundation of knowledge about and continually educating about
   - Ecological limits and disparity
   - The Ethic of Reciprocity/the "Golden Rule" and sharing
   - The paucity of knowledge and need to abide by the Precautionary
     Principle & the 2nd Law of Thermodynamics
   - Our dependence of the natural resource base and diversity

2. Realize that there are many ways to Rome, in terms of meeting Maslow's needs.

Biointensive Agriculture

Traditional Agriculture

Positively Ethical Applied Community Ecology

Hunter-Gatherer

Low Input Sustainable Agriculture

Permaculture

Organic Agriculture

Urban Agriculture

Conventional Agriculture

Deep Ecology

Minimalist Living

Pastoral

3. Practice life-long learning of ecological principles and processes.

The Starfish Parable

"Once upon a time, there was an old man who used to go to the ocean to do his writing. He had a habit of walking on the beach every morning before he began his work. Early one morning, he was walking along the shore after a big storm had passed and found the vast beach littered with starfish as far as the eye could see, stretching in both directions.

Off in the distance, the old man noticed a small boy approaching. As the boy walked, he paused every so often and as he grew closer, the man could see that he was occasionally bending down to pick up an object and throw it into the sea. The boy came closer still and the man called out, 'Good morning! May I ask what it is that you are doing?'

The young boy paused, looked up, and replied 'Throwing starfish into the ocean. The tide has washed them up onto the beach and they can't return to the sea by themselves,' the youth replied.

'When the sun gets high, they will die, unless I throw them back into the water.'

The old man replied, 'But there must be tens of thousands of starfish on this beach. I'm afraid you won't really be able to make much of a difference.'

The boy bent down, picked up yet another starfish and threw it as far as he could into the ocean. Then he turned, smiled and said, 'It made a difference to that one!'"

Adapted from "The Star Thrower", by Loren Eiseley (1907 – 1977)

https://www.goodreads.com/author/quotes/56782.Loren_Eiseley

Now, do we operate much as the boy in this parable did and put our efforts, money, resources and energy into cleaning up oil-covered birds after oil spills; rehabbing animals caught in barbed wire, hit by cars, sickened or poisoned by biocides or plastics or wounded by a poor shot; and providing "artificial" habitat and supplemental food, other artificial assistance for endangered species; etc.?

OR

Do we begin to truly and significantly change our socio-political/ economic (ecological) systems and start to live in a "*sabio*, simple, small, slow, steadfast, sharing, sustainable" way in order to provide an Earth conducive to the thriving of healthy-biodiversity of all wildlife including insects, plants, fungi,and other biota?

I think morally & ethically we do both but with an emphasis on the latter!

How? Communication in the arena of sustainability &
<u>associated</u> decision-making processes is difficult.
Parameters of time, space, ecological units, geopolitical
positions, values, ... all need to be laid out on the table and
discussed, struggled with, & assimilated by all.

> You folk really have trouble communicating, especially with the other 9 million species like us. ... You really need to work harder & smarter at it!

> We humans talk over, under, and around each other... & oftentimes do not try to listen, learn, empathize, respect, and truly communicate.

Gathering of Scientists & Academicians

NO CORRELATION

Socio/political economic, psychological walls

...Traditional home/kitchen chat of common folk

> If growing conditions are good, I generally plant by the moon ... like my dad and grandpa did.

What? We Need Ethical Energy Transformation (Photosynthesis) for
Humanity and Nature.

"Sustainably transform energy in Nature" is what **WE** do.

Ethical: A Low Input/Throughput Photosynthetic
Economy (Earth)

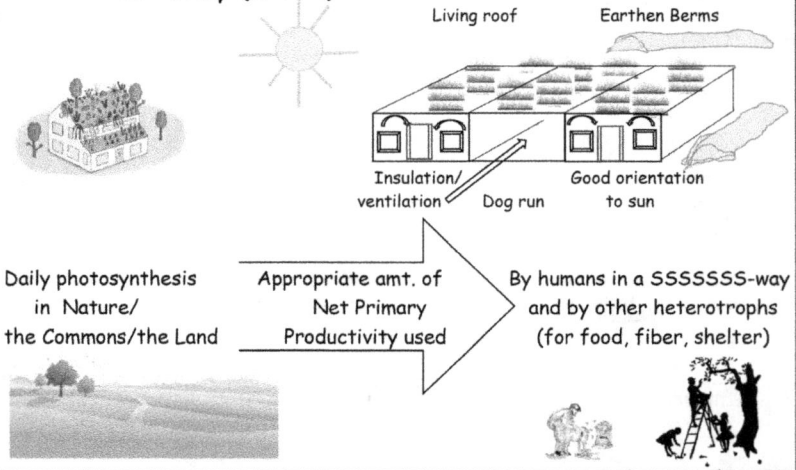

Living roof   Earthen Berms

Insulation/ventilation   Dog run   Good orientation to sun

| Daily photosynthesis in Nature/ the Commons/the Land | Appropriate amt. of Net Primary Productivity used | By humans in a SSSSSSS-way and by other heterotrophs (for food, fiber, shelter) |

Unethical: Our High Input/Throughput Fossil Energy
Economy (Eaarth/Anthropocene)

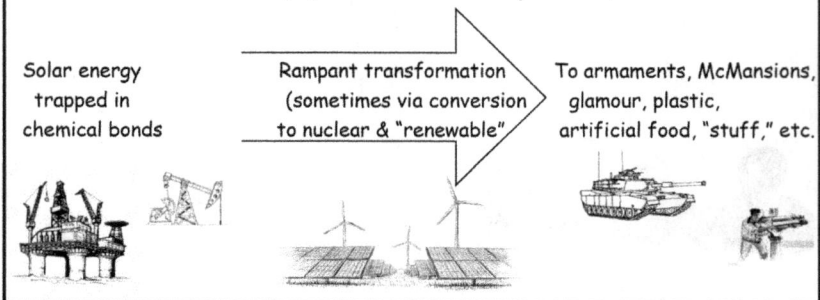

| Solar energy trapped in chemical bonds | Rampant transformation (sometimes via conversion to nuclear & "renewable") | To armaments, McMansions, glamour, plastic, artificial food, "stuff," etc. |

How? Conventional industrial agriculture, "organic" agriculture, or sustainable agriculture?

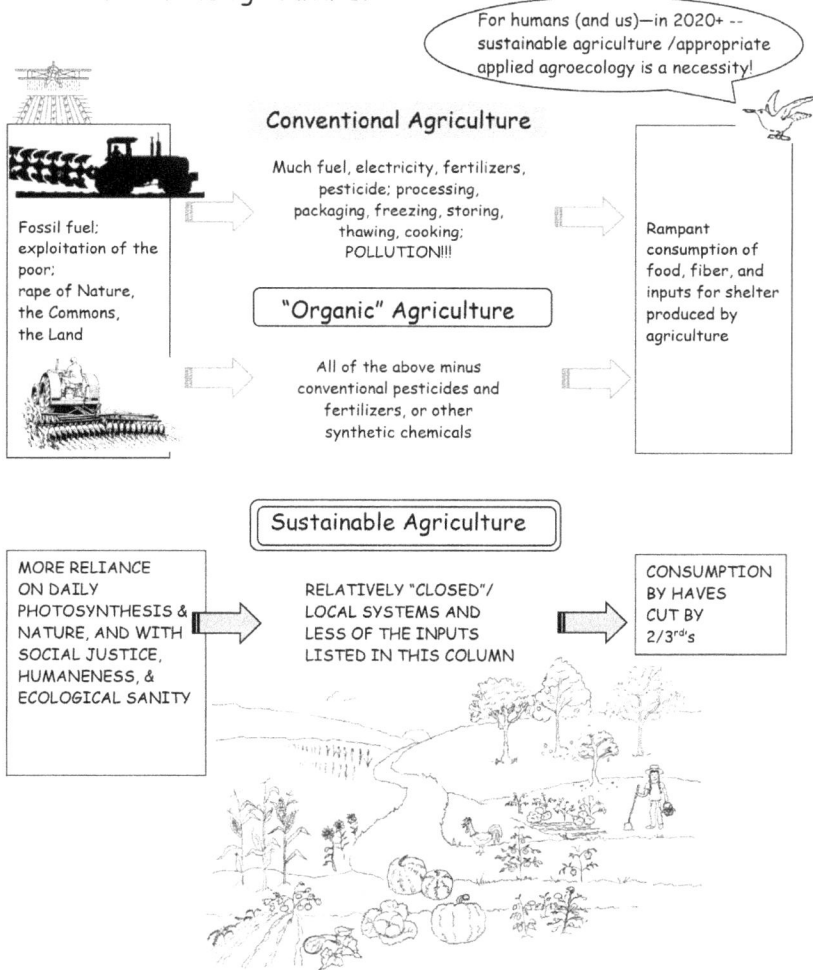

For humans (and us)—in 2020+ -- sustainable agriculture /appropriate applied agroecology is a necessity!

Conventional Agriculture

Much fuel, electricity, fertilizers, pesticide; processing, packaging, freezing, storing, thawing, cooking; POLLUTION!!!

Fossil fuel; exploitation of the poor; rape of Nature, the Commons, the Land

Rampant consumption of food, fiber, and inputs for shelter produced by agriculture

"Organic" Agriculture

All of the above minus conventional pesticides and fertilizers, or other synthetic chemicals

Sustainable Agriculture

MORE RELIANCE ON DAILY PHOTOSYNTHESIS & NATURE, AND WITH SOCIAL JUSTICE, HUMANENESS, & ECOLOGICAL SANITY

RELATIVELY "CLOSED"/ LOCAL SYSTEMS AND LESS OF THE INPUTS LISTED IN THIS COLUMN

CONSUMPTION BY HAVES CUT BY 2/3rds

<u>How?</u> By employing indicators of these following realms of sustainability with monitoring, measuring, analyzing, evaluating, assessing, re-planning, and righting the course

> If you don't have a goal with a process of measuring and assessing for sustainability, you *ain't* going to get there.

Governance—"Participatory democracy," "shared long-term vision," decision-making is open. ... Sociocracy.

"Effective management cycles, from formulation through implementation to evaluation"

Conservation of the natural resource base and reliance on daily photosynthesis in a sustainable manner. **REDUCE**, Reuse, recycle.

Strategic planning in addressing holistic community health for all (psychological, sociopolitical/economic, i.e., all ecological aspects.)

Travel by foot, horse, **bicycle**, or mass transport (preferably train, bus, or modern clipper ship)

Health and well-being for all humans and other life in community

An economy based on sustainable livelihoods/local foods and **appropriate applied agroecology**

Social equity and justice

Local to global and global to local (Commitment to assuming "global responsibility for **peace, justice**, equity, development of sustainable community and climate protection.")

> Soil organic matter, tilth?
> Soil cover with living biota? Biodiversity? Natives?
> Quantity of quality H$_2$0?
> Net primary productivity?

ARE WE THERE YET?

How? Knowledge of the interactions of matter & energy in a human-dominated world, for the good of ALL !

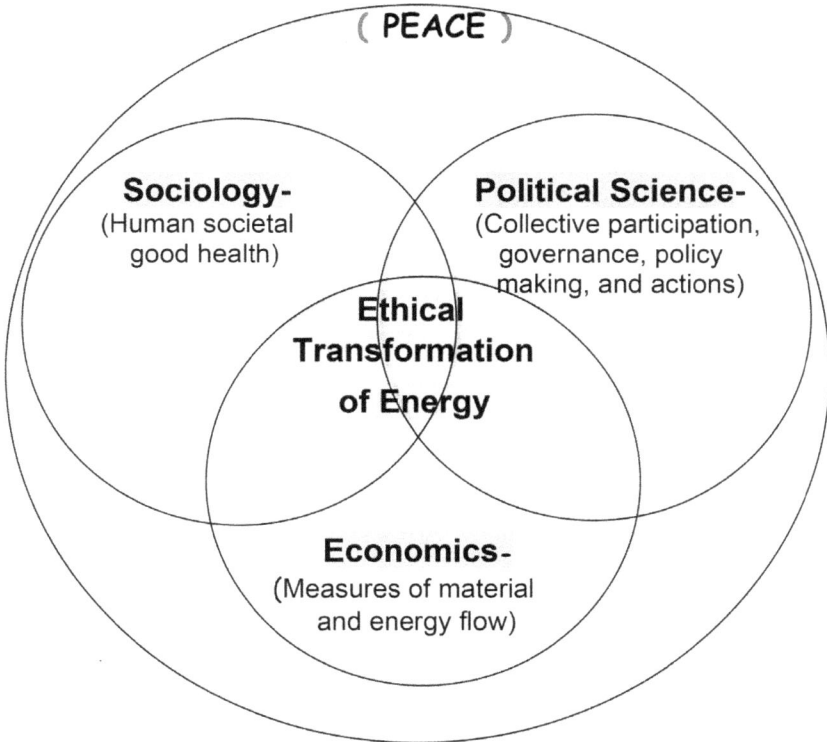

I hope you get it right!

We ARE in this together !!!

Positively Ethical Applied Community
ECOLOGY
(PEACE)

Sociology-
(Human societal
good health)

Political Science-
(Collective participation,
governance, policy
making, and actions)

**Ethical
Transformation
of Energy**

Economics-
(Measures of material
and energy flow)

Laws Which Guide Us Individually and Collectively Toward Quality Life for All[1]

Rule of Law. "It's the Law!" Natural Law. The Law. The Constitution. Neighborhood Association Laws and Regulations. Housing Codes. Motor Vehicle Law. City ordinances. Food and Drug Laws. Pesticide Law. Seed Law. Conservation District Regulations. Animal Welfare Law. Affirmative Action. Reparations. U.N. Declaration of Human Rights. The U.N. Agenda 21 Plan.
(Sometimes my head spins with this legalese and concern over associated mores.)

I strongly believe in good and more than adequate governance and laws and regulations. And, I am a conservative in terms of respecting many tried-and-true traditions and taboos, honoring the positions of elders, and of conserving

• what may exist as a fair, just, and humane social fabric,

as well as

• the natural resource base and homeostatic symbioses.

(G. Tyler Miller, Jr., who wrote many books related to applied ecology, emphasized in his "Principles of Understanding and Sustaining the Earth" that the principle of true conservatism is

"Don't ever call yourself a conservative unless what you want to conserve is the earth.")

---

1. I wrote this after several weeks of frustrations in dealing with the legal walls a desperate but kind and humble young man from Central America was facing. He and his extremely poor family and friends spent about one year, thousands of dollars, and considerable anguish and time in a process of seeking asylum in the U.S. after his life was threatened by governmental entities in Nicaragua. He has since been deported back to Nicaragua and is struggling to keep a low profile and survive. Natural law and an ethic of reciprocity should take precedence over legislative laws, executive orders, and judicial decisions or at a minimum be greatly considered in the processes of their development and implementation.

On the other hand, I also fervently believe in attempting to right the wrongs (through reparations, affirmative action, redistribution of wealth and power, "Medicare for all," "free" public education) of constitutional, legislative, and legal precedents from previous administrations and generations which have resulted in serious inequities. Some of these wrongs include colonization and intrinsic genocide of indigenous peoples, slavery, disenfranchisement, segregation and unequal education, Jim Crow sins, red-lining, unequal enforcement of laws, pervasive incarcerations, wars and subsequent land, resource, and power grabs, and other discriminatory, social unjust, inhumane, and ecologically unsound actions/practices of the past and present. This would definitely also include righting the wrongs in our "sacred" constitution(s) and in "holy" religious books no matter how long they may have been around.

However, I must admit I am not certain how I might precisely and accurately label my socio-political/economic or ecological positions. Probably democratic ecosocialist? Or maybe communitarian? Or ecological Marxist?? But certainly not morally and ethically corrupt Trumpist, conservative Republican, or right wing-libertarian, and probably not anarchist. Perhaps a leftist-libertarian socialist. (I suppose I could just play dirty pool and say that I am basically an advocate for *la mejor mezcla* [¡*No!* Not *mezcal*! Well, maybe.] of these and other political ideologies and practices which might be the best fit for desired outcomes of quality life for all for as long as possible, given the psychological, socio-economic/ecological characteristics of a local, regional, or national geographic area in terms of what can really be done and will truly fit.)

I am not a fan of free market, neoliberal capitalism. This system has disrupted the resilient and sustainable social fabric of traditional cultures and of demes, populations, and ecological communities of many other species. It has a deplorable history of being exploitive of the Commons and the Land on which sustainable cultures hunted and gathered and practiced traditional sustainable agriculture, and on which the people of these cultures lived healthy lives, which were to a large extent in concert with symbioses. In order to approach sustainability, capitalism must be severely regulated. (My biggest concern with the Trumpian wave is his pro-neoliberal capitalistic rhetoric, positions, and proposed policies which are anti-governmental regulation. However, I also do not hesitate to add that even though she was/is not so manipulatively vocal and overt about it, many of Hillary Clinton's proposed policies were also de facto neoliberal and certainly capitalistic and were not what is needed to help move us significantly from Eaarth to a neo-Earth.)

I do believe that a participatory democratic process is generally essential to an achievement of a healthier Eaarth. However, there must be checks and balances such that money, power, elitism, and might do not make right and that the tyranny of the majority does not squash the rights of minorities.

Below I present a sort of hierarchy of laws and regulations affecting and employed by human societies. Nevertheless, I am reluctant to state that one level is more important than the others. Moreover, there is considerable overlapping. I will say that the physical laws and scientific laws will ultimately rule, and that Natural Law is, in reality, the most important law concerning homeostatic symbioses (if we wish to include humans).

- Physical Laws ("Ruling" the Cosmos)
- Scientific Principles (The little we humans know that is quite solid and certain. This includes the big bang theory, the laws of thermodynamics, Newton's laws, the theory of relativity, the theory of evolution.)
- Biological & Ecological Principles (Important to us because they deal with quality life on Earth)
- Natural Law (Ethics and morals which should be involved in all the law which follows this below, and should always have precedence)
- International Law
- Federal Law/Constitutional Law (which should recognize and abide by international law in moral and ethical ways)
- State & Local Law (which includes federal/constitutional law)
- Tribal Rules
- Local Customs
- Family Traditions
- Individual Ethos

THE PROCESS OF PEACE

I appreciate young Benjamin Austin's contribution to the Wall Street Journal from Friday, March 3, 2017 about a need for an education in economics, in the more conventional sense, in high schools (*https://www. wsj.com/articles/economics-shouldnt-be-an-elective-1488498856*). However, the gigantic hole in human societies which is threatening much of life on Eaarth in the Anthropocene, including *Homo sapiens*, is ecological illiteracy. We desperately need positively ethical applied ecological economics across curricula and campuses of all human organizational entities.

Some questions related to ecological economics and sustainable community, which need to be worked on and begun to be answered for students from pre-school until death, are:

Why do we desperately need regeneration and conservation of resilient/SUSTAINABLE community?

•   Are there LIMITS concerning the natural resource base of the Earth, material flow, energetics, homeostasis, population growth, and quality life?

•   Is there disparity on Eaarth? Should we practice an ethic of reciprocity /the golden rule???? PEACE???? Love????

•   Is "Nature" (somewhat as it was) important? Are we biophiliacs??

What goals, policies, and actions are needed for regeneration and conservation of resilient/SUSTAINABLE community? Do we not need to:

- Realize positively ethical applied community ecology[1] (PEA-CE) across campuses and curricula of all human organizational entities?

- Be humble, frugal, simple, small, slow, and abide by the Precautionary Principle, realize sufficiency more than efficiency, practice sustainable livelihoods with a light ecological footprint, and leave much of "Nature" alone?

- Utilize and develop appropriate agroecology?

- Open borders; share knowledge and means toward sustainable livelihoods, Land and resources; work hard to get rid of nuclear arms, other armaments, and guns? Share toward equity/equality and respect for all life forms, races, ethnicities, genders, sexual orientations, age groups, mentally and physically challenged?

How do we go about achieving regeneration and conservation of resilient/SUSTAINABLE community? Should we

- "Lay it all out on the table" and dialogue through science and art; brainstorm; tell our stories through prose, poetry, song, music, and other arts; communicate, cogitate, ruminate, communicate? Root out even more of personal, social, cultural, political, ecological history, deal with psychological, physiological, philosophical challenges,

---

1. In order to wisely, prudently, critically think and act from birth until actually becoming an Elder, all need to be continually developing a knowledge of principles and processes in natural systems, i.e., learning about holism /connectivity, watersheds, biome characteristics and dynamics, weather & climate, evapo-transpiration, degree-day/phenology modeling and other modeling, levels of organization, the energy pyramid, "as energy is transformed/used it tends toward uselessness," trophic levels, food webs, producers and consumers, grazers, browsers, carnivores, decomposers, hydrological cycles, biogeochemical cycles, homeostasis, population /ecological community / and ecosystem dynamics, pecking orders, keystone species, limiting factors, carrying capacity, micro-/macro-evolution, small vs. big, r- & K-strategists, species interactions, allelopathy, importance of biodiversity, ecological succession, climax communities, fire ecology, soil /water /biotic community /human ecology, energetics, net primary productivity, metabolism, poikilotherms & homeotherms, least-cost/end use analysis, pollution, desertification, eutrophication, biomagnification, appropriate scale, appropriate technology, marine /aquatic/ terrestrial ecology, agroecology, "planned, controlled multispecies rotational grazing," steady-state ecological economics, How do we live well in a place?

As Aldo Leopold and David Orr said (somewhat paraphrased): all education is ecological education, and "If education does not teach us these things [of ecological mechanisms], then what is education for?"

and continue communicating?

•   Set goals, policy, action plans; take action; monitor, analyze, evaluate, assess; replan?

•   Keep on a-muddling through toward regeneration and conservation of resilient/SUSTAINABLE community?

Ecology Across Campuses and Curricula, and Ecological Literacy: Toward Sustainable Livelihoods and Conservation and Sustainable Community

From a paper presented at the 33rd Annual Conference of the Society of Educators and Scholars, October 2010.

paul b. martin et al.

https://bannedbookscafe.blogspot.com/2013/09/normal-0-false-false-false.html

We know we are in dire straits[1] as a species, and that we are among associated species facing even more difficult immediate threats, such as extinction by *Homo sapiens.* At least one billion relatively powerless humans and great numbers of individuals of other species, are in the midst of unprecedented peril at this moment. Therefore, we must demand that our institutions of learning from the pre-kindergarten to post-graduate school unreservedly address the challenge of lowering ecological footprints and material/energy usage/"abusage" (loss of topsoil, usable water, and biodiversity; and dependence on virtual slaves) in the sectors of the world with power and capable of facilitating the increases in ecological footprints for those lacking power (the objective being an average worldwide per capita ecological footprint of ca. 5 acres and daily energy usage of ca. 70 thousand kilocalories with a very small standard deviation.). Stakeholders should be basing their decisions on ecological principles and processes; they should be thinking critically & creatively and acting in local and global systems with goals of banning inequality and enhancing conservation, resilience, and sustainability. A comprehensive and intensive plan for making this happen through our educational systems is seriously needed. Development of ecology across curricula toward eco-

---

1. An excellent presentation illustrating the situation in which *Homo sapiens* finds itself as a species is at *https://www.youtube.com/watch?v=qPb_0JZ6-Rc&feature=share& fbclid=IwAR2hDnprGfWB7R1n8Z0WXhER6ZGxgZMwDh8O-mWOjYv7Fp7heR_DJB_ JYcg*

logical literacy and sustainable livelihoods and sustainable community is a moral and ethical imperative for true scholars and educators. Educational systems should be a part of the solution rather than serving to increase the social and ecological problems we all will continue to face in the future if we do not begin to rapidly change our socio-political/economic local and global systems.

Why are we not hearing more about what is fundamental to the education process and essential to critical thinking and quality life for all, i.e., ecological literacy and ecology across curricula?

We think there are at least seven fairly obvious reasons:

1. **Ecological Illiterates.** Most parents, teachers, administrators, and education policymakers in the U.S. do not have real knowledge of a concept of ecological literacy (EL) and ecology across the curriculum (EAC). Moreover, their grasp of EAC and EL is probably less than that of parent, teachers, administrators, and policymakers in, e.g., the 1930's. (The senior author's father, who had not attended college, and mother, who only went to school through the eighth grade, knew much more about ecological principles and processes than most folk coming out of college in today's world.)

A good foundational knowledge in ecological principles and processes is essential to anyone even beginning to understand socio-political/economic systems and for beginning to help move us toward correcting them in an ethical manner, i.e., for the good of rich or poor including other species and for as long as possible.

2. **Fear of Change.** When folk do have an inkling of understanding of what might be meant by EAC, they often generally want to avoid it at all costs, including the costs incurred from the sacrifice of necessary ecological knowledge and actions resulting from critical thinking which would take us toward future quality living within this ecosphere. The reason for this is that their paychecks, interest rates and dividends, yields from stocks and bonds, annuities, government checks, subsidies and assistance, i.e., their relatively comfortable conventional lifestyles, depend upon the bankrupt financial system and fragile socio-economic/political structure that is perilously propped upon a deteriorating natural resource base. And, they do not want to rock the boat!

The very powerful, in particular, are reluctant to give up power or even use what they have in order to gain increasing power which might be ethically shared with others. Moreover, many of those with power threaten those without similar power with certain job and income loss, in the event that they, the rich and powerful, should lose their own foothold on their exorbitant power.

3. **Uncompassionate Apathy.** Many people do not de facto care much about the three billion humans who really are struggling to get by in the world and we care even less about other species, especially if they are not mammals or are not relatively large or not stunningly beautiful.

4. **Difficulties in Redirecting Toward Sustainability (Particularly in a World of 7 Billion and Capitalism).** Individuals and world systems are complex, and difficult to reprogram toward conservation and sustainability. Moreover, there are many folk who settle for the status quo, and thus there are many naysayers, cynics, and con artists taking a perceived easier path, even while their actions are nudging, shoving, or leading us over the cliff.

5. **Communication Barriers.** It is extremely difficult to even begin to communicate with folk who hold a completely different system of values, especially when these values are on a compellingly attractive and even addictive (yet unfounded) "foundation."

6. **Faster Horses! Older Whiskey! Younger "Mates!" More Money!** We generally continue to worship at the altar of growth and big, fast, and noisy, and technological and artificial (and energetically and socio-economically/ecologically costly) at the expense in particular of "average" students/people.

7. **Problems in Knowing When the Well Really Is About to Run Dry.** As human populations and their appetites and their technology increase, the finite resources of this finite planet are rapidly tapped into and utilized. But to some extent, everything seems fine up until the depletion of necessary resources, especially macronutrients, essential elements, and compounds, is precariously near. It is difficult to know and predict when the last amount of life-essential resources are nearing depletion until very near the end of depletion or near the point where it is virtually energetically impossible to secure them in sufficient amounts to maintain the homeostasis of life systems. Dr. Albert Bartlett eloquently and thoroughly discussed this in 1978 in his classic "Forgotten Fundamentals of the Energy Crisis" (printed by Negative Population Growth,

April 1998) *www.npg.org/specialreports/bartlett_index.htm*

(Certainly, there have been better analyses of barriers to sustainability, e.g., *www.pambazuka.org/en/category/comment/61837*; however, the seven barriers listed are significant.)"

Fifty Years of Earth Day

Ecological Literacy and Values and Appropriate Action … or More and More Plagues?

What would an ethical and rational critical thinker and decision-maker expect?

Two hundred thousand years of living in dynamic homeostatic symbioses through a light ecological footprint.

Then from one million in 8000 BCE to one billion in 1800 to eight to ten billion soon!

Daily consumption by some of these humans rising from 20,000 kilocalories to 200,000 plus!!

Artificialized "symbioses", battered "nature," most of the Earth now already Eaarth!!![1]

YES! We'll have more plagues, erosion, droughts, desertification, hurricanes, flooding, disparity, famine, stress, abuse with synthetic chemicals and of the relatively powerless, Wars.

YES!!! More plagues.

This is the legacy we're leaving the next generations.

What else would an ethical and rational critical thinker and decision-maker expect?

We must make some difficult, systematic, systemic changes.

Hey Ross Douthat!

I did listen to your Catholic Conservative-educated, decadent words this weekend on C-SPAN.[2]

---

1. From Bill McKibben. For me the extra "a" is for agrilogistics and artificialization in the Anthropocene. *http://billmckibben.com/eaarth/eaarthbook.html*

2. *https://www.c-span.org/video/?469651-1/the-decadent-society* *https://www.youtube.com/watch?v=MR5qzmsDF50*

Yes. Ecological literacy and values are a tremendous challenge,
    and Eden is an impossibility.

Nevertheless Ross, listen to the <u>P</u>ositively<u>E</u>thical<u>A</u>pplied<u>C</u>ommu-
    nity<u>E</u>cology-makers.

Listen to Wendell, Wes, E.O., Herman, the Davids, Helmut,
    and Vaclav.[3]

Homo sapiens has over shot;

We are drawing down on topsoil, quality water, biodiversity,
    and daily solar photosynthate and the net primary produc-
    tivity on which quality human life depends.

We MUST learn and lighten our collective ecological footprint.

We gotta have True Hope and not overdo your Catholic praying
    and *mea culpas* and *gratia agos* to the Divine (including
    Donald John Trump).

We've gotta take Real and True well-planned ethical action
    through ecological knowledge and critical thinking and de-
    cision-making.

(Moreover, technological fixes won't cut it!)

Or else ...

What should an ethical, rational, critical thinker and decision
    maker expect?

---

3. Wendell Berry, Wes Jackson, E.O. Wilson, Herman Daly, David Orr, David Pimentel, Helmut Haberl, Vaclav Smil.

La piedra prieta

Creekside-Poet Truth

Tu es Petrus, et super hanc petram aedificabo eccle-
siam meam. Memento pulvis es, et in pulverem
reverteris.

We place you, dark one,
Gently into Walnut Creek.
A ceremony, a ritual.
(These human acts can be good.)

Plunge into the soothing water.
Forget about differences which pain.
Rock...
Meld as one
In time, space, all dimensions.

Object-oriented ontogeny.

Become silt or clay.
And we to carbon dioxide
And methane.
Simple organics and inorganics.

Melt into the abyss, to chaos.

Or aggregate, amalgamate, metamorphosize.

Each could possibly be something better!

Certainly not worse.

Appendices

Connected Assets of Commonwealth

(as presented by Ogallala Commons
https://ogallalacommons.org/about/commonwealth/)

The overall goal in life is **health**y individuals / demes / populations / community / ecosystems /ecosphere, involving **sense of place** and community and appropriate **spirituality**. The process should be developed on a foundation of (ecological) **education** including being informed by (ecological) **history**.

Actions in **leisure & recreation** and **arts & culture**, including the largest and most important component of "culture", the **foodshed**, should be toward appropriate maintenance of the four (4) major blocks of community/ecosystem/ecosphere: **soil & mineral cycle** and the **water cycle** (biogeochemical cycles), appropriate / resilient / sustainable transformation of **"renewable" energy** (as close as possible to daily and true use of solar energy arriving in an appropriate whole ... without significant conflict with Nature, or dynamic homeostatic symbioses), and **wildlife & the natural world** (ecological community involving succession).

Mining the Mother Lode

Andy Wilkinson

We are the tribe of the mother-lode aquifer.
Twelve hundred centuries, nomads have traveled here,
making their camps in the spring and the fall, seeking
shelter in canyons and washes and swales, building
hearths of caliche, and hunting and gathering
life that collected where water empowered it.
Even when drought plagued the prairie atop of it,
water welled-up from the sweet Ogallala lake
all along Yellowhouse Draw to the canyonland,
nourishing passersby, nomad and animal,
nourishing all who tread lightly and carefully.

Here in the land of the mother-lode aquifer,
rain's unpredictable, even in good seasons,
never enough, but for grasses and buffalo,
never enough, but for seasonal wanderers,
never enough for the dwellings of permanence
needed for farming and ranching and industry,
never enough for the chambers of commerce. Rain
can't be entrusted to God and the elements,
not by the tribe of the mother-lode aquifer.

Deep in the earth through the rock that encumbers it,
down to the water sand, down to the water pay,
dig down with drilling rigs, lay in the well casing,
thrust in the sucker-rod, pull it out, let it come
drawing the water up; drive it with wind-power,
drive it with gasoline, drive it electrically,

pumping and pumping and pumping 'til water runs
shining in furrows and sparkling on summer lawns,
spewing through towers for cooling the gas-flaring,
coal-smoking power plants making more energy
pumping more water, more water, more water, all
over the land of the mother-lode aquifer.

Here are no headwaters, little replenishing
what we are draining, so little restraining how
much we are using and how we are using it,
here the great lake of the Plains subterranean
dwindles each season, each turn of the faucet, each
flick of the switch starting up the submersibles,
dwindling down ditches through siphon tubes, dwindling
        down
side-rolls and pivots and gated pipe, dwindling down
water gaps, water mains, water taps, water drains,
dwindling down every new housing development,
dwindling until there are farms metamorphosing
once-irrigated to dry-land and grass pasture,
letting their silos stand empty as metaphor,
testament, future-shock here in the present-tense
frailty, the fragile, the mother-lode aquifer.

Humbling enough is this waste of our own making;
here, where we once believed rain followed plow, believed
boosters, promoters, and huckster developers,
hitched-up our wagons to forty small acres, plowed
fence-row to fence-row with cash crops on bank notes, built
churches, raised children and sent them to colleges,
sent them to wars, sent them out of the hinterlands,
send them to places that never relinquished them.
Here, from the land of the mother-lode aquifer,
people are leaving for jobs in the popular
cities, are leaving as victims of bottom-line
corporate discounters driving off businesses

started by yours and my mom-and-pop grandparents,
corporate farmers replacing the families,
swashbucklers, slashing and cutting, efficiency
chanted as mantra, while nobody's answering
who will take care of the mother-lode aquifer?

Fear lines our pocket-books, fear comes in quarter-inch
four-by-eight plywood sheets nailed over window panes,
fear grows in weeds in the sidewalks of vacancies,
fear breeds the desperate bargaining: jobs! bring us
jobs! bring us jobs! bring us jobs! bring us anything,
bring us the worst of your wastes and your prisoners,
radioactive and toxic, the detritus,
social and otherwise, flushed from the gutter-pipes
laid from the centers of power and influence,
aimed at the weak, at the people of choicelessness,
stumbling around in the wastes not their own making,
wastes that will poison the mother-lode aquifer.

Humbling enough is this come-hither beggaring,
pleading, abasing ourselves with our appetites;
worse, still, the mother-lode aquifer's guardians
shockingly favoring selling our water rights,
falling to pitches from old-fashioned renegades
nowadays using computers for running-irons,
nowadays using their lawyers for wire-cutters,
nowadays throwing out sound-bites for lariats,
bullying water boards into considering
selling our lifeblood at low bid, not worrying
selling tomorrow to pay for today, selling
every last drop of the mother-lode aquifer.

What will become of us when we are waterless?
we of the tribe of the mother-lode aquifer,
nomads and wanderers rooted by water wells,
cities and homesteads and farmlands and cattle spreads,

everything other than short grass and buffalo
wholly dependent on mining the mother-lode?
Far away, far away, where rain is plentiful
year-in and year-out and always predictable,
learned professors have studied the exodus
made by our people, our water, our resources,
calling our depopulation a certainty,
saying why fight it? let's recognize lost causes
when they are lost causes, let's give the prairie back,
back to the ruminants, back to the grasses, let's
give us a home where the buffalo roam, where the
skies are not cloudy all day after day after
day after day where the antelope seldom are
heard for there's no one to hear the discouraging
word when the commons belong to the buffalo -
crazy! say chambers of commerce, but who's crazy
now, as we drink up our mother-lode aquifer?
now, as we poison our mother-lode aquifer?
now, as we sell-off our mother-lode aquifer?

Poets and dreamers, the only true realists,
live in the future, they do not imagine it,
seeing tomorrow with yesterday's sorrowings,
seeing tomorrow as here-and-now's borrowings,
seeing the present as future's own history,
Poets and dreamers, the only true realists,
know that the gift is the ultimate mystery,
knowing a gift not in motion is powerless,
knowing no gift can be taken for profiting,
knowing no gift can be subject to ownership.
Poets and dreamers who live on El Llano know
what is the gift but the mother-lode aquifer?

What will we do with this gift of the mother-lode?
Pray that the poets and dreamers remember it,
pray this its guardians hold it in stewardship,

pray that we honor it, pray that we husband it,
pray for the tribe of the mother-lode aquifer,
pray for the water, the sweet Ogallala lake,
nourishing all who tread lightly and carefully,
lightly and carefully, lightly and carefully.

Andy Wilkinson © 2002

Enrolada[1]

paul bain martin
Jan 2007

Depression and Pinchot-wired parents taught us to
   conserve.
2007 Chosen People-fundamentalist preachers impel us to
   grab and rule the all that WE MOST CERTAINLY deserve.

Humility, frugality used to be preached in the churches.
America First!!! War!! *Estados Unidos sempre precisa*
   ganhar! ... And the ship of state lurches.

Donald Trump, Warren Buffett, Bill Gates. The power reigns.
Others squalid in big city/rural remains.

So green! So green! A new gas-electric hybrid auto and a
   muito verde LEED home built for only 200,000 times
   what half the individuals in the world make in a year.
"*Organicissimo!! Organicissimo*!!" With this certified
   pesticide-free apple sent from New Zealand for our
   fresh fruit Texas party of Christmas cheer.

"Environmentalists"' answer is "Renewable." Used only a
   half million BTU's per capita last year of ethanol and
   fuel cell energy.
...Did I mention it cost about a million British Thermal Units
   of oil, gas and coal. Don't you love the synergy?!?!

---

1. Brazilian Portuguese (slang) for "all balled up or totally messed up!!" or "complicated", often associated with procrastination.

Cars cause super-problems; we rush for more.
9-11! Pres says rush for the store.

Indoor pools, treadmill exercise machines..
Could we become Mexican Raramuri or Kenyan Kalenjins
     and make the whole earth Nature's park and run it
     in our jeans … and get it back into concert with our
     genes?

Hot outside? Turn on the *arcondicionados*! Keep eating
     more gas, oil, coal fueled potatoes.
Global warming (Climate change)? Highly recommend
     moving from Barbados.

The Chosen few of the world possess Capital and are
     Landed,
While the Third World is kwashiorkor- stranded.

7 billion *Homo sapiens*—Number ONE!!!
Other top trophic species? You're certainly done!

Top soil? Watersheds? Diverse biotic communities?
     Passive solar energy? Who the hell has heard of "it??"
Big screens, NFL, NBA, empty calories—chips and beer.
     All of this is what keeps the U.S. lit.

Farm subsidies, conservation easements, cheap energy and
     basic resources from abroad. For the Lorded Landed
     this spells more cash!
Certainly can't cap excessive income and capital gains, strive
     for real equity and equality and produce less trash!

Trans Texas Corridors! Loops around Interstate jams.
     Muckity, muckity, muck, muck, muck.
I'll solve it by traversing over Land with my on-steroids
     Hummer, ATV, Suburban, and Big diesel truck.

Frequent flyer miles, ecotourism, luxury 'beachcombing,"
    Carnival Cruise.
Explain for me again now those South American
    favela blues?!

Come to our High Schools and Universities and learn to
    change the world faster.
No matter, this "serves" to make the Natural ecosphere
    much less of a laster.

Mold those young 'uns into businessmen, corporate
    lawyers, sports physicians, oil field geologists.
But, Heaven forbid, an environmentalist or world-renowned
    ecologist!

It's the parents' fault and the teachers' *culpa*, the
    administrators get in the way. A federal problem, a
    local one. The village! The family! The individual
    student! ... Let's get more realistic!!!
It's all of these. Just start somewhere and stay focused on
    it with lots of will and energy. Get off your butt and
    think holistic!

Radios, TV's, cell phones, iPods, video games, plastics,
    petroleum perfumes, Taco Bell food, this damn
    computer screen.   Peggy Lee cries out, "Is That All
    There Is?" ... VIRTUAL reality?
Think this is really messing with my Natural, innate
    personality!

Go out into the country, inner city, or even suburban streets.
In this wealthiest of nations you'll find unattended obesity,
    high blood pressure, diabetes, asthma, cancer,
    malnutrition-- to which I would attribute bad (or lack
    of) local doctors, politics and economics, chambers of

commerce, ecological ignorance and apathy, and high
fructose corn-based sweets.

Arteriosclerosis, arthrosclerosis, hypertension, embolism,
angina, arrhythmia, heart attack, stroke. Could it be
our way of life??
Perhaps we could just slow down and do it the natural way
and give (human) Nature less strife??

Small is beautiful! Don't let them tease you!!
Big is passé... *pasado. Communidades que pensam grandão*
are screwed.

War. Basic/Airborne Ranger/Green Beret Special Forces
training into fit muscled/artificial "Army of One"- MEN
(and woMEN). Uniforms, weapons, order, brass gives
us meaning.
Could we all do a chaotic Peace Corps thing, and rather than
destroy, do just cleaning?

Haifa, Chechnya, Darfur, Cuba, the deep dark Congo.
Their notoriety? "Isn't one of those where they invented
the bongo?"
....
I'm confused!!!!
Or too much BS infused?

Got to act local.
I'm not just a Seguin yokel.

Rollin' in Sweet Earth's Arms[1]

"Rollin' in my sweet baby's arms
Rollin' in my sweet baby's arms
Lay around the shack
Till the mail train comes back
I'm rollin' in my sweet baby's arms"

Wish I's rollin' in sweet Earth's arms
Just a rollin' in sweet Earth's arms
Gotta get off my back
Ain't 'nuff good-railroad-track
Rollin' in sweet Earth's arms

Well ... Beautiful spaaacious skies
Beautiful spaaacious skies
They were there
For the natives and the bear
Beautiful spaaacious skies

You think that this is your Land
But once, believe me, 'twas mine
"You pushed my nation
To the reservation"[2]
You stole all this Land. Such a crime

---

1. I was listening to Leon's version of "Roll in My Sweet Baby's Arms" on Mattson Rainer's 92.1 FM New Braunfels on a Sunday afternoon on the way to Devine down 35 when much of this started rollin' out. On this voyage (in our old pickup) to care for Mom for a couple of days, I frantically and dangerously scribbled and captured the essence of the verses on a note pad which was luckily in my suitcase in the adjacent seat.

2. From Pete Seeger's version of Woody's "This Land Is Your Land".

Don't want amber waves of grain
No not those amber waves of grain
They're imported you see
And killed man-y a tree
These damn amber waves of grain

Hey! …
Don't you love … the warm sunshine
Frio water that really is so fine
South Texas clean air
Native grasses over there
Wish this truly was our sweet lair

Gotta reduce our ec'logical footprint
Remove that bad synthetic scent
Yes, I'm telling you true
Or this Earth'll be all through
Gotta reduce our ec'logical footprint

Artificial can be a terrible menace
In agreement with me is old Dennis
Cars, big homes should be banned
Do more with our legs and hands!
Artificial can be a terrible menace.

Living within our means
Living within our means
That does not mean
Brand-spanking new jeans
Living within our means

Well, you ask what's the 'cautionary Principle
Well, this is the best that I can do--
Critically think

Or you're thick as two short planks
That's the 'cautionary Rule

Social justice is the most important
Must all heed... the Golden Rule
But ecological sanity
Is also good for humanity
Holistically... heed the Golden Rule

Yeah, it really is a big disgrace
I'll tell you right to your face
Unions are an ace
But Cit' United, a disgrace
I will tell you right to your face

Agin!...
Cit' United just ain't quite right
No! Cit' United just ain't quite right
Was all about
Lousy quests for more might
Cit' United just ain't quite right

He's no statesman, haaas no shame
No decency ne'er takes the blame
I'm so god-damned depressed
As you may well have guessed
Ugly America's the name of his game

Winning without any ethics
Lying and brainwashing, too
They really piss me off
Wanna dump 'um in a pig trough
Winning without any ethics

It all has to do with complacency
Brought on by religions, you see

Some propaganda to win
Even screw your next of kin
It's mostly the gotdamn apathy

The only thing of wooorth, you see
LBJ said it... to both you and me
Teach', preach', pooolitician
Is what you gotta be
The only thing of wooorth, you see

Natural resources drawn doown an' overshot
It really really makes me hot
(Global Climate Change)
Nevertheless
There's one thing that is worse
Disparity is such a dreadful curse

We've all gotta share with all others
Leave some for the poor and forlorn
For the Palestinians over there
For all the birds, butterflies and hare
Leave some for the poor and forlorn

Peace is what we must strive for
It's that for which we do long
Pacifism!
Yes, nonviolence!!
Got to be very, very strong

Hey! ...
There's one thing I have to tell you!
And tell ol' Jack and Judy too
If this isn't a hit
I don't give a shit
Making War is rotten through and through.

No!...
I don't care what they think about me
I really don't give a damn
Got to try for PEACE[3]
Our lives are but on lease
Yeah, I really don't give a damn

Yeah! ...
I'm not the best of examples
No, I'm not a thing for you to be
Still I have to try
For PEACE,[3] or I will cry
Though I'm not the best of examples.

Well! ...
Ain't going to hell or to heaven
Most of me not going to the sky
I'll be down in the earth
Where I'll wish you less dearth
Hope to be with some PEACE[3] before I die.

"Rollin' in my sweet baby's arms
Rollin' in my sweet baby's arms
Lay around the shack
Till the mail train comes back
I'm rollin' in my sweet baby's arms"

Wish I's rollin' in sweet Earth's arms
Just a rollin' in sweet Earth's arms
Gotta get off my back
Ain't 'nuff good-railroad-track
Rollin' in sweet Earth's arms

---

3. Positively Ethical Applied Community Ecology.

Hope?

(November 7, 2020)

We can have some real hope
For young ones and those to come ...

When we baptize them
And cleanse them of our original sins,
Sins against indigenous peoples and slaves,
Sins against those of color ...
When we baptize them with robust local and global
Affirmative action and reparations,

When we throw out the money lenders who provide the
    capital
For the military-industrial complex, nuclear arms, other
    armaments (local and global),
And Wars,

When we rid ourselves of capital for ecological destruction.

I'll have hope when we all begin learning
Ecological principles, processes, and values
And begin to realize the rights of health care for all in the
world ...
And sustainable livelihoods
Which are truly sustainable for all, ...
Other species, the poor, ... and the rich.

Joe and Kamala,

It will be tough.

You'll have to be smart, work hard, and sacrifice much.
In solidarity!
"Please!" do give us some real reasons for hope.
Not for MAGA empires
With the lion's share of the natural resource base
And power over Nature and the poor,
But for a U.S.A., a European Union, a China, a Russia, o
        Brasil, an Israel, a Mexico,
A Korea, a Philippines, an Egypt, a Hungary, a Venezuela,
        a Honduras, a Nicaragua,
An Iran, a Belarus, the Ukraine, the Congo, a Nigeria,
A Syria, a Yemen, an Afghanistan, a Somalia, an Iraq ...
For Trump's s-hole countries, for all our s-hole countries

Which will truly address collectively, transnationally,
In solidarity,
Social injustice, inhumanness, and ecological insanity.

Give us hope for a world without the power of billionaires,
A world which empowers the truly poor and relatively
        powerless
With physical and mental well-being, with Maslow's needs,
Profound and holistic ecological health,
Concerted efforts toward scientific knowledge, wisdom, and
        prudence
And in route to critical thinking and appropriate
        decision-making
With transnational practice of the golden rule, the
        precautionary principle,
And abiding by the Second Law of Thermodynamics
As well as the United Nations Universal Declaration of
        Human Rights
And Agenda 21.

Give us a world that respects Gandhism rather than
        Trumpism,

A world of individual lives and collective-connected life
Living in solidarity
Sabiamente, simply, smally, slowly, steadfastly, sharingly,
Sustainably.
 We say to those to the mighty right,
"Your might, or wishes to be so, do not make right.
It is a sin. A mortal sin. A deadly sin. An 'unforgiveable' sin!"

To those on the left who point to and vehemently criticize
        the grave sins in the past,
Including the recent past
Of the long and exhaustive, billionaire-moneyed campaign
Of yours, Joe and Kamala, ...
We say to our dear sisters and brothers on the left who are
        quick to point fingers,
"Joe and Kamala are relatively good people close (or closer)
        in values
To the truly good people of demos."
Forgive them (and others to the left, AND the right, AND
        of the middle)
For their original, venial, mortal, deadly, "unforgivable" sins
And work towards their rehabilitation.
Force and support the appropriate and best use of their
        new powers.

Let's all try, as hopeless as it might seem
To all lighten our ecological footprints,
Our energy transformation,
Our embodied appropriation of net primary productivity.
Let us all try to get along on this third rock from the sun--
On which we are de facto stuck--
Through actions affirmative and recent Paris Accords,
And even more robust justice system reforms than Trump's
        incredibly good one,
Nuclear disarmament treaties and defunding our War
        Department,

The banning of other arms,
Defunding and disarming police
And getting them out of schools
In real efforts toward curing the root causes of social
        disease and disharmony.
Let us work toward regeneration and conservation and de-
velopment
Of resilient, sustainable ecological community.

Let's work toward realizing sustainable livelihoods in s-hole
        countries,
All of our respective s-hole countries here on Eaarth.

May we work to have <u>Faith</u> in ecological science,
To be charitable and to genuinely Love one another and all
of Nature,
And realize real <u>Hope</u>.

Let's all let our hearts bleed liberally for justice,
Hug some trees, and
Attempt to commune peacefully,
As positively ethical applied community ecologists,
For the good of the future progeny of *Homo sapiens*
        and other species,
For the good of Nature ... dynamic homeostatic symbioses.

The challenges are great.
But the goals of Bernie and AOC and others of like minds
Are mostly good and their efforts admirable,
And they are achievable with appropriate and persistent
        strategic planning and action,
Effective monitoring, assessment, and replanning to right a
        course.

Joe and Kamala,

You all and Speaker Pelosi do have Mitch McConnell et al.,
The very red states, and the right-wing Supreme Court with
     whom to deal.
Still ... keep your hopes up and persevere! They all, we all,
     have SOME decency!!!

Some say our socio-ecological challenges are
     insurmountable.
Others posit that Capitalism based on the original sins
Of genocide of and Land grabs from indigenous peoples
And exploitation of Nature and relatively powerless humans
Is the best system possible
(Mostly, I think, because they are Haves who Have!).
Others say significant change from the current exploitive
     system,
Though needed,
Would be against their traditions, their tribe, their family,
The values of their spouse, significant others, loved ones.
They want their kids to have more
Even when their kids have more than enough more
And there isn't enough for all kids, including kids of other
     species, to have more.
Perhaps loved ones don't deserve to be loved so much??
Or rather, perhaps true love, though it may be perceived as
     callous tough love,
Is facilitating a learning toward living--and dying--in concert
     with all other life?

Seven to eight billion humans (going on TEN!)
And their substantially more numerous domesticated
     species
(Utilitarian species as well as sometimes only decorative
     house cats and dogs)
Artificialize and cover most of the globe.
And even though we've left little for wild species,
Many "humans" seem to want more

Which will inevitably eliminate these beautiful, natural
    creations of the wild.
More wanting more,
Worshipping at the altars of money and stuff and recreation.
Re-creating as if they were God
Rather than respecting and living in concert
With the real god, dynamic homeostatic symbioses.
Plowing, mowing, leaf-blowing
With millions of fossil fuel-driven machines,
Concreting, asphalting, plasticizing, building many relatively
    useless structures,
Trains of 18-wheelers and automobiles continuously
    traveling up and down the interstates
When real trains would be so much better ecologically,
Because laissez-faire capitalism demands it!
Hard-wiring of Earth with cables across land and sea.
Soft -wiring ubiquitously, currently with the latest
    capitalistic technology, 5G, across the airs
Because neo-liberal capitalism demands it!

Do I have hope.
Yes. ...
Some hope.

¿¡Esperanza!?
Cuando empezamos a vivir sabiamente ...
Sabiamente, simply, smally, slowly, steadfastly, sharingly,
Sustainably.
In solidarity.

Little Things We Can Do To Have Healthy And Better Life Along With A Healthier, More Sustainable Earth[1]

Eat healthy and exercise regularly. Truly integrate this into your workday and lifestyle.

Experience the real world—your yard, open fields, farms, ranches, parks—and stay away from TV, computers, cell phones, video games and other electronic gadgetry. Get physically, intellectually, and spiritually in touch with Nature, the Land, Community and People (the very young and Elders).

For short trips: walk, run, ride bikes, or skateboard.

For long trips: car-pool. or take a train or bus.

Help your family start a garden, maybe an organic garden.

Appropriately incorporate grazers, browsers, and other heterotrophs into local rural/urban foodsheds.

Volunteer to help with community gardens or your school's garden.

Buy something at the Farmers' Market and get to know the farmers.

Encourage your family to mulch-mow and to mow, water & fertilize the yard less. Use locally adapted native vegetation and introduce vegetable and fruit-producing garden areas into the landscape.

Compost all leaves, grass clippings, food scraps, and other organic matter.

Buy less, reuse neat old things, and recycle. Carry a cup for drinking.

Use less!!! And use less plastic. Do not use disposable (plastic) water bottles.

Keep air conditioning and heating systems off. Open windows. (At

---

1. Upon request by my wife, Elizabeth Florence Martin, I (paul bain martin) originally developed this list one Earth Day morning many years ago for Betsy's science classes at Seguin High School.

least keep thermostats low in the winter and high in the summer.)

Help caulk cracks around windows, doors, and in other leaky areas of your home. Place weather-stripping around doors.

Use less water. Take shorter showers, catch water & and use in the sink, help family fix dripping faucets, etc.

Put a bucket/tub in your shower to collect the "warm up" water and overspray. Use it in your garden.

Use rain barrels to catch roof run-off. Your plants will love the soft, low mineral water.

Completely turn off lights/electricity users! Use power strips!

At stores refuse plastic bags. Take your own homemade "cool" bag.

Hang your clothes to dry out on a line and let the Sun/wind do the job.

Build a solar-water heater. Do not use an electric water heater.

Prepare to be an educated and responsible ecological-friendly voter who is active in local and global community.

Work/have a career in "jobs" that help others & enhance ecological systems. (Sustainable livelihood)

Work on a farm, ranch, summer camp and/or park system for the summer and/or after school.

Learn about ecological, carbon, water, and energy "foot-printing," & life-cycle analysis.

Encourage peers and adults to be truly responsible in using prescription drugs, alcohol, etc. ... Work hard at discouraging addictive drugs such as nicotine (smoking cigarettes, snuff/smokeless tobacco, vaping, etc.) and other such drugs.

Learn about your family and community/regional history.

Learn about the flora and fauna of your backyard, nearby vacant lot, and local community. (Natural History).

Go for a walk with a small child and teach him/her/etc. the names of birds, other animals, trees, other plants, mushrooms, and other biota (cyanobacteria, bacterial disease symptoms) that you see along the way.

Read, write, and do arts and crafts of some type. Do "hands on" projects and keep your mind challenged with mathematical and other puzzles, problem-solving, and critical & creative thinking.

Frequently write and call your local, state, and federal political representatives about appropriate policy concerning positively ethical applied community ecology.

Conserve, help the poor with "hands up" and stay debt-free.

Work for open borders and real enduring Peace and Justice. (War is NOT the answer!)

Live *"Sabiamente*, Simply, Smally, Slowly, Steadfastly, Sharingly… SUSTAINABLY

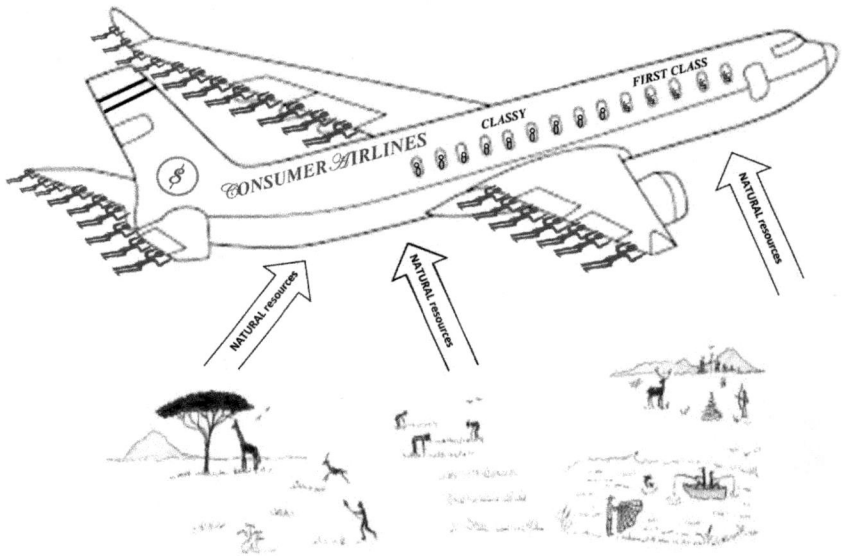

The plane is labeled "CONSUMER AIRLINES" with seating sections "FIRST CLASS" and "CLASSY". Arrows labeled "NATURAL resources" point upward from the landscape scenes toward the plane.

Christmas Message 2019

I begin this christmas message with a theme which often comes forth in my mind. The image therein is a metaphor for today's world, a 737 Max at 35,000 feet of altitude with about 0.5-1 billion super-Haves living relatively comfortably (though extremely artificially and unsustainably) in the first-class section of the cabin, 3 billion in economy, and 3 billion relative have-nots clinging to the outer shell of the plane. There are Nature-exploiting tubes extending from down on relative paradise (Earth/Eaarth) supplying necessary fuel for the plane and the inhabitants, and some of the fuel entering the cabin does leak out and trickle down as crumbs to those clinging onto the exterior. The plane cannot land because (1) the pilots are absent or inept, (2) a very poor navigation system exists, and (3) non-functional landing gear is in use, i.e., lack of ecological knowledge and values and critical thinking and decision-making. There are a few in and around the plane working on the navigation and landing systems; however, most in the cabin are just busy living the "good" life and/or working hard to keep the plane flying (with the military/police, fossil fuel providers, artificial information/algorithm-systems personnel being the most important), and those outside are both working to keep the plane flying as well as struggling to get inside the doomed international capitalistic system. Down below on the rest of Eaarth, there are a very few humans and other species living very comfortable low-input/-throughput (sort of Paleolithic or early Neolithic) lives in concert with what is left of dynamic homeostatic symbioses (from the perspective of humans and species associated with quality life for humans).

The Eaarth has seven (7) billion humans going on ten (ten) and a considerable amount of energy devoted to military/police, fossil fuel, and the maintenance of flight of the human economy. Moreover, mindsets, attitudes, and behavior are relatively rigid and not conducive to developing a sustainable world. Therefore, means for getting out of this predicament are extremely challenging. However, such an effort would start with:

• a learning system of "positively ethical applied community ecology/PEACE" across all human organization entities,

• a more holistic, comprehensive, and profound understanding of ecological principles and processes especially connectivity, the Second Law of Thermodynamics, the Precautionary Principle, net primary productivity and embodied human-appropriated net primary productivity, and an ethic of reciprocity and pacifism, and

• a development of plans and actions in solidarity from those perspectives with a goal of: light ecological human footprints, individually and collectively, and increase in parity within the human population as well as realizing protection and enhancement of habitat for other species.

Four Changes

by Gary Snyder

Herein are excerpts from Bioneers website, *https://bioneers. org/four-changes-by-gary-snyder/?fbclid=IwAR09QLhHmyUZwTCkYy- qD8Mak61Yq2P7Kco1GCqhG1G04dpuvfThjbMuTxYw* :

"In July 2016, Jack Loeffler recorded Gary Snyder reading his updated version of 'Four Changes' in his home. This recorded version was prepared for and included in a major exhibition held at the History Museum of New Mexico at the Palace of the Governors in Santa Fe.

The exhibition was entitled 'Voices of Counterculture in the Southwest', and Snyder's rendering of 'Four Changes' aptly conveyed how deeply the counterculture movement helped nurture the emerging environmental movement. The impact of this manifesto is as powerful today as it was a half century ago and could not be more timely.

Four Changes at Age 50: A Celebration on the Environmental Movement's First Manifesto of Contemplative Ecology. Introduction by Diana Hadley, Jack Loeffler, Gary Paul Nabhan and Jack Shoemaker.

In the months before the first Earth Day in April 1970, mention of a prophetic manifesto seemed to crop up in nearly every serious discussion of what the nascent environmental movement should be and what values it should embody. That manifesto was conceived and shaped in the summer 1969, as poet Gary Snyder toured a number of college campuses around the United States and then entered into deeper discussions with a number of other poets, visionaries and activists in the San Francisco Bay area. Affectionately called 'Chof' by other radical environmentalists during that time, Snyder gradually refined their collective vision into a ten page draft document that became what we now know as *Four Changes*.

Several features of this manifesto were then, and still are, unique in the canon of writings considered foundational to the environmental movement. Snyder's literary gifts shine through the manifesto with prescient, poetic and playfully comic qualities to them. The tone seemed as fresh and as 'out of the box' as *Leaves of Grass* must have sounded when Whitman first sowed it onto the American earth a century earlier. The manifesto called for a radical shift in our relationship with the planet through changing the way we perceive *population, pollution, consumption*, and the *transformation* of our society and ourselves. In this manner, it foreshadowed later expressions of *ecological thought* that we now call *contemplative ecology* and *deep ecology*.

While it was in many ways anchored in Buddhist teachings, it was also precise in its understanding of modern ecological science and respectful of the place-based wisdom of the traditional ecological knowledge of the many indigenous cultures of the world. It did not privilege Western science over other ways of making sense of the environment, but welcomed dialogue and integration of many distinctive expressions.

Four Changes was also rooted in a mature understanding of the political ecology of power dynamics and disparities in access to resources that were ravaging our planet, its biological and cultural diversity. Parts of it were so pertinent to these issues that it was read into the *Congressional Record* on April 5th, 1970— two and a half weeks before Earth Day flags were unfurled all around the world. In that sense, it was perhaps the first robust articulation of what we now call a yearning for *environmental justice*. Still, the tone was hopeful—that humankind could learn to respect, learn from and embrace the other-than-human-world. As Snyder later paraphrased one of the tenets of *Four Changes*, '*Revolutionary consciousness is to be found among the most ruthlessly exploited classes: animals, trees, water, air, grasses.*' It is time to heed the call of the prophetic *Four Changes*."

"POPULATION: THE CONDITION

Position: Human beings are but a part of the fabric of life — dependent on the whole fabric for their very existence. As the most highly developed tool-using animal, we must recognize that the unknown evolutionary destinies of other life forms are to be respected, and we must act as gentle steward of the Earth's community of being.

Situation: There are now too many human beings, and the problem is growing rapidly worse. It is potentially disastrous not only for the human race but for most other life forms.

Goal: The goal would be half of the present world population, or less."

"POLLUTION: THE CONDITION

Position: Pollution is of two types. One sort results from an excess of some fairly ordinary substance—smoke, or solid waste—that cannot be absorbed or transmuted rapidly enough to offset its introduction into the environment, thus causing changes the great cycle is not prepared for. (All organisms have wastes and by-products, and these are indeed part of the total biosphere: energy is passed along the line, refracted in various ways. This is cycling, not pollution.) The other sort is powerful modern chemicals and poisons, products of recent technology that the biosphere is totally unprepared for. Such are DDT and similar chlorinated hydrocarbons—nuclear testing fallout and nuclear waste—poison gas, germ and virus storage and leakage by the military; and chemicals that are put into food, whose long-range effects on human begins have not been properly tested.

Situation: The human race in the last century has allowed its production and scattering of wastes, by-products, and various chemicals to become excessive. Pollution is directly harming life on the planet: which is to say, ruining the environment for humanity itself. We are fouling our air and water, and living in noise and filth that no 'animal' would tolerate, while advertising and politicians try to tell us 'we've never had it so good.' The dependence of modern governments on this kind of untruth leads to shameful mind-pollution through the mass media and much school education.

Goal: Clean air, clean clear-running rivers, the presence of Pelican and Osprey and Gray Whale in our lives; salmon and trout in our streams; unmuddied language and good dreams."

"CONSUMPTION: THE CONDITION

Position: Everybody that lives eats food and is food in turn. This complicated animal, the human being, rests on a vast and delicate pyramid of energy transformation. To grossly use more than you need to destroy is biologically unsound. Much of the production and consumption of modern society is not necessary or conducive to spiritual and cultural growth, let alone survival; and is behind much greed and envy, age-old causes of social and international discord.

Situation: Humanity's careless use of 'resources' and its total dependence on certain substances such as fossil fuels (which are being exhausted, slowly but certainly) are having harmful effects on all the other members of the life-network. The complexity of modern technology renders whole populations vulnerable to the deadly consequences of the loss of any one key resource. Instead of independence we have over-dependence on life-giving substances such as water, which we squander. Many species of animals and birds have become extinct in the service of fashion fads — or fertilizer — or industrial oil. The soil is being used up; in fact, mankind has become a locust-like blight on the planet that will leave a bare cupboard for its own children — all the while in a kind of Addict's Dream of affluence, comfort, eternal progress — using the great achievements of science to produce software and swill.

Goal: Balance, harmony, humility, growth that is a mutual growth with Redwood and Quail — to be a good member of the great community of living creatures. True affluence is not needing anything.

ACTION

Social/Political: It must be demonstrated ceaselessly that a continually 'growing economy' is no longer healthy, but a cancer. And that the criminal waste which is allowed in the name of competition — especially that ultimate in wasteful needless competition, hot wars and cold wars with 'communism' (or 'capitalism') — must be halted totally with ferocious energy and decision. Economics must be seen as a small sub-branch of Ecology, and production/distribution/consumption handled by companies or unions or cooperatives with the same elegance and spareness one sees in nature. Soil banks; open space; logging to be truly based on sustained yield (the US Forest Service is sadly now the lackey of business). Protection for all predators and varmints. 'Support your right to arm bears.' Damn the International Whaling Commission which is selling out the last of our precious, wise whales! Ban absolutely all further development of roads and concessions in National Parks and Wilderness Areas; build auto campgrounds in the least desirable areas.

Initiate consumer boycotts of dishonest and unnecessary products. Establish Co-ops. Politically, blast both 'Communist' and 'Capitalist' myths of progress, and all crude notions of conquering or controlling nature.

The Community: Sharing and creating. The inherent aptness of communal life — where large tools are owned jointly and used efficiently. The power of renunciation: If enough Americans refused to buy a new car for one given year it would permanently alter the American economy. Recycling clothes and equipment. Support handicrafts — gardening, home skills, midwifery, herbs — all the things that can make us independent, beautiful and whole. Learn to break the habit of acquiring unnecessary possessions, a monkey on everybody's back — but avoid a self-abnegating anti-joyous self-righteousness. Simplicity is light, care-free, neat, and loving — not a self-punishing ascetic trip.

(The great Chinese poet Tu Fu said, 'The ideas of a poet should be noble and simple.') Don't shoot a deer if you don't know how to use all the meat and preserve that which you can't eat, to tan the hide and use the leather — to use it all, with gratitude, right down to the sinew and hooves. Simplicity and mindfulness in diet are the starting point for many people."

"TRANSFORMATION: THE CONDITION

Position: Everyone is the result of four forces — the conditions of this known-universe (matter/energy forms, and ceaseless change); the biology of his or her species; individual genetic heritage; and the culture one is born into. Within this web of forces there are certain spaces and loops that allow to some persons the experience of inner freedom and illumination. The gradual exploration of some of these spaces constitutes 'evolution' and, for human cultures, what 'history' could increasingly be. We have it within our deepest powers not only to change our 'selves' but to change our culture. If humans are to remain on Earth they must transform the five-millennia-long urbanizing civilization tradition into a new ecologically-sensitive, harmony-oriented, wild-minded scientific/spiritual culture. 'Wildness is the state of complete awareness. That's why we need it.'

Situation: civilization, which has made us so successful a species, has overshot itself and now threatens us with its inertia. There is also some evidence that civilized life isn't good for the human gene pool. To achieve the changes, we must change the very foundations of our society and our minds.

Goal: nothing short of total transformation will do much good. What we envision is a planet on which the human population lives harmoniously and dynamically by employing various sophisticated and unobtrusive technologies in a world environment that is 'left natural.' Specific points in this vision:

A healthy and spare population of all races, much less in number than today.

Cultural and individual pluralism, unified by a type of world tribal council. Division by natural and cultural boundaries rather than arbitrary political boundaries.

A technology of communication, education, and quiet transportation, land-use being sensitive to the properties of each region. Allowing, thus, the Bison to return to much of the high plains. Careful but intensive agriculture in the great alluvial valleys; deserts left wild for those who would live there by skill. Computer technicians who run the plant part of the year and walk along with the Elk in their migrations during the rest.

A basic cultural outlook and social organization that inhibits power and property-seeking while encouraging exploration and challenge in things like music, meditation, mathematics, mountaineering, magic, and all other ways of authentic being-in-the-world.

Women totally free and equal. A new kind of family — responsible, but more festive and relaxed is implicit.

ACTION

Social/Political: It seems evident that there are throughout the world certain social and religious forces that have worked through history toward an ecologically and culturally enlightened state of affairs. Let these be encouraged: Gnostics, hip Marxists, Teilhard de Chardin Catholics, Druids, Taoists, Biologists, Witches, Yogins, Bhikkus, Quakers, Sufis, Tibetans, Zens, Shaman, Bushmen, American Indians, Polynesians, Anarchists, Alchemists . . . the list is long. Primitive cultures, communal and ashram movements, cooperative ventures.

Since it doesn't seem practical or even desirable to think that direct bloody force will achieve much, it would be best to consider this change a continuing 'revolution of consciousness' which will be won not by guns but by seizing the key images, myths, archetypes, eschatologies, and ecstasies so that life won't seem worth living unless one's on the side of the transforming energy. We must take over 'science and technology' and release its real possibilities and powers in the service of this planet — which, after all, produced us and it. More concretely, no transformation without our feet on the ground.

Stewardship means, for most of us, find your place on the planet, dig in, and take responsibility from there. The tiresome but tangible work of school boards, county supervisors, local foresters, local politics, even while holding in mind the largest scale of potential change. Get a sense of workable territory. Learn about it and start acting point by point. On all levels, from national to local, the need to move toward steady state economy, equilibrium, dynamic balance, inner growth stressed must be taught – maturity, diversity, climax, creativity.

The Community: New schools, new classes, walking in the woods and cleaning up the streets. Find psychological techniques for creating an awareness of 'self' that includes the social and natural environment. 'Consideration of what specific language forms — symbolic systems — and social institutions constitute obstacles to ecological awareness.' Without falling into facile interpretations of McLuhan, we can hope to use the media. Let no one be ignorant of the facts of biology and related disciplines; bring up our children as part of the wildlife. Some communities can establish themselves in backwater rural areas and flourish — others maintain themselves in urban centers, and the two types work together — a two-way flow of experience, people, money, and home-grown vegetables. Ultimately cities may exist only as joyous tribal gatherings and fairs, to dissolve after a few weeks. Investigating new lifestyles is our work, as is the exploration of ways to explore our inner realms — with the known dangers of crashing that go with such. Master the archaic and the primitive as models of basic nature-related cultures — as well as the most imaginative extensions of science — and build a community where these two vectors cross.

Our Own Heads: are where it starts. Knowing that we are the first human beings in history to have so much of our past cultures and previous experiences available to our study, and being free enough of the weight of traditional cultures to seek out a larger identity – the first members of a civilized society since the early Neolithic to wish to look

247

clearly into the eyes of the wild and see our selfhood there, our family there. We have these advantages to set off the obvious disadvantages of being as screwed up as we are — which gives us a fair chance to penetrate some of the riddles of ourselves and the universe, and to go beyond the idea of 'human survival' or 'survival of the biosphere' and to draw our strength from the realization that at the heart of things is some kind of serene and ecstatic process that is beyond qualities and beyond birth and death. 'No need to survive! In the fires that destroy the universe at the end of the kalpa, what survives?' — 'The iron tree blooms in the void.' Knowing that nothing need be done is the place from which we begin to move."

Good Books Dealing with Important Aspects of Positively Ethical Applied Community Ecology/ PEACE

Altieri, Miguel. 1995. *Agroecology: The Science of Sustainable Agriculture, 2nd Ed.* **Sustainable food systems are in concert with Nature and ethical science in contrast to industrial agriculture which attempts to dominate nature.** *https://monthlyreview.org/2009/07/01/agroecology-small-farms-and-food-sovereignty/*

Armstrong, Karen. 1993. *A History of God.* **Karen helps us to understand humanity!** *https://www.theguardian.com/books/2009/jul/04/case-for-god-karen-armstrong*

Berry, Wendell. 2003. *The Art of Commonplace: the Agrarian Essays of Wendell Berry.* **Berry is a wonderful thinker who is truly the master naturalist/positively ethical applied community ecologist!!** *http://blog.spu.edu/signposts/book-review-the-art-of-the-commonplace/*

Bohlen, P.J. & Gar House. 2009. *Sustainable Agroecosystem Management: Integrating Ecology, Economics, and Society.* **We are applied ecologists in whole interacting systems.** *http://books.google.com/books/about/Sustainable_Agroecosystem_Management.html?id=ohrCU0lIxIEC*

Brimlow, R.W. 2006. *What About Hitler? Wrestling with Jesus's Call to Nonviolence in an Evil World.* **War begets War. Pacifism, nonviolence, diplomacy, satyagraha, peaceful protest, civil disobedience, economic and political noncooperation are true routes to PEACE!** *https://www.publishersweekly.com/978-1-58743-065-7*

Brody, Hugh, 2000. *The Other Side of Eden: Hunters, Farmers, and the Shaping of the World.* **This is a terrific book by this anthropologist! It provides some good insights into how to be an ethical applied ecologist—in a holistic fashion**. *http://www.theguardian.com/books/2001/jan/28/society*

Brown, Peter. 2015. *Ecological Economics for the Anthropocene: An Emerging Paradigm.* **Peter is working hard and intelligently to straighten the whole mess out.** *https://www.youtube.com/watch?v=cTCf4glFzxM*

Brooker, R.J. et al. 2017. *Principles of Biology, 2nd Ed.* **We all need to have a solid understanding of life systems in order to be good citizens and voters!** *https://cbs.umn.edu/contacts/robert-brooker*

Catton, William. 1982. *Overshoot: The Ecological Basis of Revolutionary Change.* **Drop into this wonderful book to see what condition our condition is in.** *http://www.ncbi.nlm.nih.gov/pmc/articles/PMC2602943/*

Coates, Peter. 1998. *Nature: Western Attitudes Since Ancient Times.* **Another must-read for naturalists.** *https://www.goodreads.com/book/show/1380770.Nature*

Conrad, Jessica. 2014. *Sharing Revolution: The essential economics of the commons.* **It is by enhancing the assets of commonwealth that we will realize quality life for all. (There are several recent good books discussing this commons topic and its import to reality.)** *http://onthe-commons.org/magazine/our-new-ebook-sharing-revolution*

Corbett, J. & M. Corbett. 2000. *Designing Sustainable Communities: Learning from Village Homes.* **The Corbetts present some good thoughts, concepts, strategies and tactics.** *http://digitalcommons.calpoly.edu/cgi/viewcontent.cgi?article=1001&context=crp_fac*

Cox, G.W. & M.D. Atkins. 1979. *Agricultural Ecology: An Analysis of World Food Production Systems.* **This is an oldy but a goody to use to begin to peek into agroecology.** *http://books.google.com/books/about/Agricultural_Ecology.html?id=wmxoQgAACAAJ*

Cronon, William. 1991. *Nature's Metropolis. Chicago and the Great West.* **It takes a lot from Nature to build the artificial.** *https://www.nytimes.com/2011/03/28/opinion/28krugman.html*

Crosby, Alfred W. 2004. *Ecological Imperialism. The Biological Expansion of Europe 900-1900, 2nd Ed.* **There was rapid changing of the biosphere with European expansion.** *https://en.wikipedia.org/wiki/Ecological_Imperialism_(book)*

Daly, Herman E. 1991. *Steady-State Economics, 2nd Ed.* 1996. *Beyond Growth: The Economics of Sustainable Development.* **Daly proposes a sustainable ecological economic world-system.** *https://eebweb.arizona.edu/courses/Ecol206/DalyHermanSteady-StateEconomics.pdf*

Daly, Herman E.; Cobb, John B., Jr. 1994. *For the Common Good: Redirecting the Economy toward Community, the Environment, and a Sustainable Future, 2nd Ed.* **This book addresses the why for ecological economics!** *http://www.ecobooks.com/books/comgood.htm*

Doughty, Robin W. 1983. *Wildlife and Man in Texas: Environmental Change and Conservation.* **Read this and other books by Doughty and get insights into what Texas looked like in the 1800s, etc.** *http://www.utexas.edu/cola/depts/geography/faculty/rdoughty*

Galeano, Eduardo. 2013. *Children of the Days. A Calendar of Human History.* **Everyone should have a book by Galeano in their library!** *https://www.nytimes.com/2013/07/28/books/review/children-of-the-days-by-eduardo-galeano.html*

Goldman, Daniel. 2009. *Ecological Intelligence: How Knowing the Hidden Impacts of What We Buy Can Change Everything.* **After life cycle/whole systems analysis of energetics/material flow/etc. we often find that:**

- renewable energies are not so renewable,

- appropriate is inappropriate,

- natural is unnatural, and

- that green is not green.

http://www.danielgoleman.info/topics/ecological-intelligence/

Grandin, Greg. 2019. *The End of the Myth. From the Frontier to the Border Wall in the Mind of America.* **A fantastic book dealing with Manifest Destiny, constant wars, financial meltdown, and reactionary populism and racist nationalism.** *https://www.thenation.com/article/greg-grandin-end-of-the-myth-frontier-border-wall-book-review/*

Haberl, Helmut et al. 2016. *Social Ecology: Society-Nature Relations Across Time and Space.* **This book covers the current state of the art in social ecology as researched by the Vienna School of Social Ecology!** *https://www2.hu-berlin.de/iri-thesys/refbase/files/jonasonielsen/2017/305_JonasO.Nielsen2017.pdf*

Haidt, Jonathan. 2012. *The Righteous Mind. Why Good People Are Divided by Politics and Religion.* **This is a good up-to-date book on human behaviors.** *https://www.nytimes.com/2012/03/25/books/review/the-righteous-mind-by-jonathan-haidt.html*

Hickel, Jason. 2020. *Less Is More: How Degrowth Will Save the World*. **This book is one of the best! It covers the mess in which we are, how we got there, and with great hope, how we will get out of it.** *https://www.resilience.org/stories/2020-10-05/less-is-more-how-degrowth-will-save-the-world-by-jason-hickel/*

Jackson, Wes. 1992. *Becoming Native to This Place*. **A great little quick-read which can facilitate our process of becoming naturalists.** *https://centerforneweconomics.org/publications/becoming-native-to-this-place/*

Isenberg, Andrew. 2000. *The Destruction of the Bison*. **A wonderful book which is very historically informative and reveals mis-/dis-information about the amazing animal we call the buffalo and the peoples associated with it.** *https://www.researchgate.net/publication/300039885_Review_of_The_Destruction_of_the_Bison_An_Environmental_History_1750-1920_By_Andrew_C_Isenberg*

Kirschenmann, Frederick. 2010. *Cultivating an Ecological Conscience: Essays from a Farmer Philosopher*. **Fred is a leader in moving us toward sustainable agriculture, agroecology, and positively ethical applied community ecology/PEACE.** *http://muse.jhu.edu/books/9780813173733*

Klein, Naomi. 2007. *The Shock Doctrine. The Rise of Disaster Capitalism*. **The destruction of Nature and much of the good of the social fabric by neoliberal capitalism and economic policies of privatization, deregulation, and cuts to social services has been realized as a result of force, stealth and crisis.** *https://www.nytimes.com/2007/09/30/books/review/Stiglitz-t.html*

Loflin, B., S. Loflin, & S.L. Hatch. 2006. *Grasses of the Texas Hill Country*. **There are many good plant identification books/websites which can be helpful in this process. ... This book is excellent with good photos!** *https://www.tamupress.com/book/9781585444670/grasses-of-the-texas-hill-country/*

Mann, Charles. 2005. *1491: New Revelations of the Americas Before Columbus*. **In this well-written book, Mann covers recent and intriguing archaeological/anthropological information on the Americas/the World before Columbus' "discovery."** *http://www.nytimes.com/2005/10/09/books/review/09baker.html?_r=0*

Mann, Charles. 2011. *1493: Uncovering the New World Columbus Created*. **This is not as well-written as 1491 but nevertheless is a must-read for recent archaeological/anthropological information** on the world after Columbus. *http://www.nytimes.com/2011/08/21/books/review/1493-uncovering-the-new-world-columbus-created-by-charles-c-mann-book-review.html?pagewanted=all*

McDaniel, Carl. 2005. *Wisdom for a Livable Planet.* **This is about some very significant leaders in positively ethical applied community ecology/PEACE.** *http://tupress.org/books/wisdom-for-a-livable-planet*

Miller, Char. 2001. *On the Border: An Environmental History of San Antonio.* **Miller is a prolific author of research works on environmental history!!** *https://www.goodreads.com/author/list/8103.Char_Miller*

Miller, G. Tyler, Jr. 1990. *Resource Conservation and Management.* **This is an old textbook, but a very good book for a starter into PEACE!** *https://en.wikipedia.org/wiki/Natural_resource_management*

Moerman, Daniel. 1998. *Native American Ethnobotany.* **This big book contains much valuable information on plants and their uses.** *http://books.google.com/books/about/Native_American_Ethnobotany.html?id=U-XaQat5icHUC*

Nash, Roderick Frazier. 1990. *American Environmentalism: Readings in Conservation History.* **This is a good history of positively ethical applied community ecology/PEACE.** *http://books.google.com/books/about/American_environmentalism.html?id=rGJRAAAAMAAJ*

Norton, Bryan. 1998. *Toward Unity Among Environmentalists.* **Norton has helped us understand the history/policy leading to efforts of positively ethical applied community ecology and the influences of Pinchot and Muir.** *http://www.jstor.org/discover/10.2307/30301526?uid=3739920&uid=2129&uid=2&uid=70&uid=4&uid=3739256&sid=21103886007461*

Odum, H.T. & E.C, Odum. 2004. *The Prosperous Way Down.* **Drs. Howard Odum and David Pimentel helped some of us to think energetically. "Our civilization can thrive in a future where we live with less."** *http://prosperouswaydown.com/*

Orr, David. 1992. *Ecological Literacy: Education and the Transition to a Postmodern World.* **This is a very good book by the educator who wrote the classic paper, "What Is Education For?"** *http://books.google.com/books/about/Ecological_Literacy.html?id=iKJbrKO9TwgC*

Pimentel, David. 1980. *Handbook of Energy Utilization in Agriculture.* **Dr. Pimentel led the way in having a better understanding of the energetics of industrial and sustainable agriculture!** *https://rodaleinstitu-*

te.org/organic-pioneer-dr-david-pimentel/

Prashad, Vijay. 2012. *The Poorer Nations: A Possible History of the Global South.* **Until we Haves understand how we seriously exploit the people south of the Tropic of Cancer through our conventional life-styles and until we understand how these life-styles detrimentally affect other species directly and indirectly, we can never claim to be true naturalists or positively ethical applied community ecologists.** *https://blogs.lse.ac.uk/lsereviewofbooks/2013/03/28/book-review-the-poorer-nations-prashad/*

Real, L.A. & Brown, J.H. 1991. *Foundations of Ecology: Classic Papers with Commentaries.* **Papers which ecologists have been introduced to in their education.** *http://press.uchicago.edu/ucp/books/book/chicago/F/bo3613618.html*

Reiman, J. & P. Leighton. 2010. *The Rich Get Richer and the Poor Get Prison: Ideology, Class and Criminal Justice.* **To cling onto Nature we all must be naturalists/positively ethical applied community ecologists. But it is difficult to be a naturalist in prison. "Natural" isn't natural unless we all are able to appreciate and benefit from Nature, and the poor, including those in prison, have just as much right to be master naturalists as the rich and maybe more. (If you cannot take in and digest this, remember that in ecology, all is connected—and connected in what are very complex systems.)** *http://books.google.com/books/about/The_rich_get_richer_and_the_poor_get_pri.html?id=CvjtAAAAMAAJ*

Reynolds, H.L. et al. 2010. *Teaching Environmental Literacy: Across Campus and Across the Curriculum.* **If we are going to make much headway toward positively ethical applied community ecology/PEACE, we are going to have to realize ecology across curricula/campuses:** *http://bannedbookscafe.blogspot.com/2013/09/normal-0-false-false-false.html*

Sapolsky, R. M. 2017. *Behave: The Biology of Humans at our Best and Worst.* **This book brings us up to date in terms of neurological and psychological understanding of human behavior.** *https://www.nytimes.com/2017/07/06/books/review/behave-robert-m-sapolsky-.html*

Savory, A. 2016. *Holistic Management, Third Edition: A Commonsense Revolution to Restore Our Environment.* **For the collaborators on this little applied ecology book, Savory's diagram of a holistic thought model for decision-making has been his most important contribution. However, perhaps he has gotten too balled up in the planned-controlled, rotation-grazing possibilities for appropriate "MANagement"!**

https://www.context.org/iclib/ic25/wood/

Schmidly, David. 2002. *Texas Natural History: A Century of Change.* **Nature is dynamic!!** *https://www.goodreads.com/book/show/2211379.Texas_Natural_History*

Schumacher, E.F. 1973. *Small Is Beautiful.* **This is truly a CLASSIC! It is largely why my mantra is** *Sabio***, Simple, Small, Slow, Steadfast, Sharing, Sustainable.** pbm *https://en.wikipedia.org/wiki/Small_Is_Beautiful*

Smil, Vaclav. 2013. *Harvesting the Biosphere: What We Have Taken From Nature.* 2017. *Energy in Nature and Society.* **Any naturalist/ecologist/biologist worth his or her salt must read Smil!** *http://www.vaclavsmil.com/category/books/*

Stone, M.K. & Z. Barlow. 2005. *Ecological Literacy: Educating Our Children for a Sustainable World.* **We must teach children and their immediate and extended family and friends about how to realize quality life for all including other species for as long as possible, and how to be satisfied with needs instead of wants.** *https://www.sustainable.org/living/education-training-and-lifelong-learning/126-ecological-literacy-educating-our-children-for-a-sustainable-world*

Thoreau, Henry David. 1862. *Walking.* **"In short, all good things are wild and free." And at this URL is the whole book:** *https://www.theatlantic.com/magazine/archive/1862/06/walking/304674/*

Turner, M.W. 2009. *Remarkable Plants of Texas.* **This is an informative and fun book!** *http://utpress.utexas.edu/index.php/books/turrem*

Vitek, Bill & Wes Jackson. 2008 *Virtues of Ignorance: Complexity, Sustainability, and the Limits of Knowledge.* **We need to be much less hubristic and recognize our limitations to truly think and act critically and sanely.** *http://books.google.com/books/about/The_Virtues_of_Ignorance.html?id=2-8BaV-_LpUC*

Weatherford, Jack. 1988. *Indian Givers: How the Indians of the Americas Transformed the World.* **This easy-read reveals contributions to ecology and democracy from Native Americans.** *http://www.jstor.org/discover/10.2307/1185432?uid=3739920&uid=2129&uid=2&uid=70&uid=4&uid=3739256&sid=21103885493491*

White, Richard. 2011. *Railroaded. The Transcontinentals and the Making of Modern America.* **The government and big corporate businessmen have been a major player in America's rugged individualis-**

tic destruction of Nature and Native American genocide. *https://www.
nytimes.com/2011/07/17/books/review/book-review-railroaded-by-richard-white.
html*

Wilson, E.O. 1992. *The Diversity of Life.* 2014. *The Meaning of
Human Existence.* **These are just two of many great books by this wonderful ant-man at Harvard.** *http://en.wikipedia.org/wiki/Consilience_%-
28book%29 http://www.paperbackswap.com/Edward-O-Wilson/author/*

Wilson, E.O. 2016. *Half-Earth: Our Planet's Fight for Life.* **E.O.
tells why and how we need to support Nature.** *https://eowilsonfoundation.
org/half-earth-our-planet-s-fight-for-life/*

Worster, Donald. 1994. *Nature's Economy: A History of Ecological
Ideas, 2nd Ed.* **This book should be read by all naturalists.** *https://blogs.
commons.georgetown.edu/cs525/2009/10/21/natures-economy-donald-worster/*

The following URL takes developing PEACE-makers to many
other classic books dealing with positively ethical applied community
ecology/PEACE: *https://www.goodreads.com/shelf/show/ecology* Also, *The
Economist* magazine, *The NYT, The BBC, NPR, PBS, The New Yorker* do
give a very good daily/weekly overviews of the state-of-the-states of the
world.

Cartoons Which Help in a Powerful Way to Tell the Story of PEACE

1.Communicating the Truth.

Two persons walking down the street with one pensively saying, *"I can't decide if we're good people who are bad at communicating, or monsters who communicate perfectly."*

L. Finck, The New Yorker 10/9/17

https://condenaststore.com/featured/i-cant-decide-if-we-are-good-people-who-are-bad-at-communicating-liana-finck.html

Two cute mutt dogs walking down the street and one grumbles, *"I bark and I bark but I never feel like I effect real change."*

C. Weyant, The New Yorker 10/15/18

https://condenaststore.com/featured/i-bark-and-i-bark-christopher-weyant. html

2. Eaarth.

A barefoot and distraught man crawling in rags across a clear-cut landscape comes to a sign (with a vulture sitting on it) which reads: *"Pardon our dust … as we continue to remodel every square inch of plant earth! Humankind Inc., A Pan-Global Organization. 'We're everywhere you are!'"*

B. Veley, Population Connection 12/2018

https://www.cartoonstock.com/cartoonview.asp?catref=bven1605

3. Sustainability (Including "Sufficiency & Efficiency").

At the saloon bar one rough cowboy with a small fedora-type hat challenges one with a big ten gallon-type, *"Well stranger, I reckon you'd be welcome to stay if you replaced that ten-gallon hat with an efficient low-flow model.' Showdown in Sustainable Gulch."*

Wayno & Piraro, Bizarro 7/19/19

https://www.bizarro.com/cartoons

An older yuppie man reaching to store a reusable water container, opens a large cabinet packed with similar permanent plastic and aluminum water containers, i.e., stuffed to the point that some fall out on the floor when he opens it. His significant other exclaims, *"At what point does it stop being sustainable?"*

M. Larson, The New Yorker 10/28/19

https://www.newyorker.com/cartoon/a23019

4. Global Climate Change.

A fellow outside in a burning and terribly hot landscape pleas through a window *"I said, this has been the hottest decade in history!"* … A "Republican?" elephant inside next to a window replies, *"What?? I can't hear you when the air conditioner is so loud!"*

Tom Toles, Washington Post 2020

https://www.washingtonpost.com/people/tom-toles/

5. Silly, Artificial, Technological-Fix Shell Games (Appropriate Technology?).

Corporate business **dialogue about a new invention for removing global-climate change excesses of CO₂** from the atmosphere—which trees need for energy capture … **and whether people have the scientific expertise and rationality to stop the removal of CO2 at the right moment** in order to have appropriate photosynthesis/NPP and to achieve dynamic homeostatic symbioses (get us back to "nature" and Earth).

Scott Adams, Dilbert 2/3/19

https://dilbert.com/strip/2019-02-03

6. Positively Ethical Applied Community Ecologists?

A very long line queued up at an information kiosk for folk *"preocupados com o futuro do meio ambiente"* or **"worried about the future of ecological systems"** with no one queued up at a kiosk for folk *"dispostos* *a mudar o estilo de vida*" or **"prepared to change their lifestyle to address ecological challenges"**.

ó menina, Próximo via ComuniDária 8/25/19

https://callparaasparedes.blogs.sapo.pt/proximo-298288

7. Empathy?

A psychologist advisedly says to her very disappointed CEO patient, **"Many CEOs suffer from bouts of empathy toward workers. Reading your compensation agreement at bedtime should quell those thoughts."**

Wayno & Piraro, Bizarro 12/27/18

https://www.bizarro.com/cartoons

8. "Green" Technologies? Sound Energetics?

An electric car plugged into a huge coal plant with the caption, **"How Electric Cars Work."**

M. Ramirez, Student Daily News 3/16/2012

https://www.studentnewsdaily.com/editorial-cartoon-for-students/how-electric-cars-work/

Thor is battling three bad-a… aliens with his lightning-powered hammer. Two funky-appearing bystanders are looking on, one wearing a **"Stop CO2"** t-shirt and the other with a **"Save the Planet"** sign. The fellow with the t-shirt says, *"It's fine that you can save us from aliens, Thor. But how much energy does that thingy use? And how about fixing the climate?"*

M. Wulff & A. Morgenthaler Wumo 4/19/2020

https://www.arcamax.com/thefunnies/wumo/s-2350703

www.ingramcontent.com/pod-product-compliance
Lightning Source LLC
Chambersburg PA
CBHW060351200326
41519CB00011BA/2108